TWENTY FIRST CENTURY
Science

Project Directors

Angela Hall
Emma Palmer
Robin Millar
Mary Whitehouse

Authors

Philippa Gardom Hulme
Frank Sochacki

OXFORD
UNIVERSITY PRESS

Great Clarendon Street, Oxford OX2 6DP

Oxford University Press is a department of the University of Oxford.
It furthers the University's objective of excellence in research,
scholarship, and education by publishing worldwide in

Oxford New York

Auckland Cape Town Dar es Salaam Hong Kong Karachi
Kuala Lumpur Madrid Melbourne Mexico City Nairobi
New Delhi Shanghai Taipei Toronto

With offices in
Argentina Austria Brazil Chile Czech Republic France Greece
Guatemala Hungary Italy Japan Poland Portugal Singapore
South Korea Switzerland Thailand Turkey Ukraine Vietnam

Oxford is a registered trade mark of Oxford University Press
in the UK and in certain other countries.

First published 2012.

British Library Cataloguing in Publication Data.

Data available.

ISBN 978-0-19-913835-7

10 9 8 7 6 5 4 3 2

Printed in Great Britain by Bell and Bain Ltd, Glasgow.

Paper used in the production of this book is a natural, recyclable product
made from wood grown in sustainable forests. The manufacturing process
conforms to the environmental regulations of the country of origin.

Acknowledgements
Illustrations by IFA Design, Plymouth, UK, Clive Goodyer, and Q2A Media.

Author acknowledgements
Many thanks to Catherine and Sarah for checking the puzzles, and to
Barney for his inspirational ideas. Thanks to Ruth for her careful editing,
and to Les, Sophie, and Barry at OUP for all their help and patience.

About this book

Welcome to the Twenty-First Century Biology Revision Guide! This book will help you prepare for all your GCSE Biology module tests. There is one section for each of the biology modules B1–B7, as well as six sections covering Ideas about science. Each section includes several types of pages to help you revise.

Workout: These are to help you find out what you can remember already, and get you thinking about the topic. They include puzzles, flow charts, and lots of other types of questions. Work through these on your own or with a friend, and write your answers in the book. If you get stuck, look in the Factbank. The index will help you find what you need. Check your answers in the back of the book.

Factbank: The Factbanks summarise information from the module in just a few pages. For B1–B7, the Factbanks are divided into short sections, each linked to different statements in the Specification. The Ideas about science Factbanks are different. They are conversations, covering the ideas you will need to apply in different contexts. Read them aloud with a friend if you want to.

Quickfire: Sections B1–B7 each have Quickfire questions. These are short questions that cover most of the content of the module. For some questions, there is space to answer in the book. For others, you will need to use paper or an exercise book.

GCSE-style questions: These are like the questions in the module tests. You could work through them using the Factbank to check things as you go, or do them under test conditions. The answers are in the back of the book. Each section for B1–B7 has one 6-mark question, designed to test your ability to organise ideas, and write in clear and correct English. Use these to help you practise for this type of question in the module tests.

H In every section, content covered at Higher-tier only is shown like this.

Other help: This page and the next one include vital revision tips and hints to help you work out what questions are telling you to do. Don't skip these!

Making the most of revision

Remember, remember: You probably won't remember much if you just read this book. Here are some suggestions to help you revise effectively.

Plan your time: Work out how many days there are before your test. Then make a timetable so you know which topics to revise when. Include some time off.

Revise actively, don't just read the Factbanks. Highlight key points, scribble extra details in the margin or on Post-it notes, and make up ways to help you remember things. The messier the Factbanks are by the time you take your tests, the better!

Mind maps: try making mind maps to summarise the information in a Factbank. Start with an important idea in the middle. Use arrows to link this to key facts, examples, and other science ideas.

Test yourself on key facts and ideas. Use the Quickfire sections in this book, or get a friend to ask you questions. You could make revision cards, too. Write a question on one side, and the answer on the other. Then test yourself.

Try making up songs or rhymes to help you remember things. You could make up **mnemonics**, too, like this one for the classification hierarchy:

Kids **P**refer **C**hocolate **O**ver **F**ancy **G**reen **S**alad

Apply your knowledge: Don't forget you will need to apply knowledge to different contexts, and evaluate data and opinions. The GCSE-style questions in this book give lots of opportunities to practise these skills. Your teacher may give you past test papers, too.

Ideas about science: should not be ignored. These are vital. In your module tests, there could be questions on any of the Ideas about science you have covered so far, set in the context of most of the topics you have covered.

Take short breaks: take plenty of breaks during revision – about 10 minutes an hour works for most people. It's best not to sit still and relax in your breaks – go for a walk, or do some sport. You'll be surprised at what you can remember when you come back, and at how much fresher your brain feels!

Answering exam questions

Read the question carefully, and find the command word. Then look carefully at the information in the question, and at any data. How will they help you answer the question? Use the number of answer lines and the number of marks to help you work out how much detail the examiner wants.

Then write your answer. Make it easy for the examiner to read and understand. If a number needs units, don't forget to include them.

Six-mark questions

Follow the steps below to gain the full six marks:

- Work out exactly what the question is asking.
- Jot down key words to help your answer.
- Organise the key words. You might need to group them into advantages and disadvantages, or sequence them to describe a series of steps.
- Write your answer. Use the organised key words to help.
- Check and correct your spelling, punctuation, and grammar.

Below are examiner's comments on two answers to the question: ***"Describe the benefits to your body of a good diet and explain the need to balance diet and exercise."***

✎ The quality of written communication will be assessed.

Command words

Calculate Work out a number. Use your calculator if you like. You may need to use an equation.
Compare Write about the ways in which two things are the same, and how they are different.
Describe Write a detailed answer that covers what happens, when it happens, and where it happens. Your answer must include facts, or characteristics.
Discuss Write about the issues, giving arguments for and against something, or showing the difference between ideas, opinions, and facts.
Estimate Suggest a rough value, without doing a complete calculation. Use your science knowledge to suggest a sensible answer.
Explain Write a detailed answer that says how and why things happen. Give mechanisms and reasons.
Evaluate You will be given some facts, data, or an article. Write about these, and give your own conclusion or opinion on them.
Justify Give some evidence or an explanation to tell the examiner why you gave an answer.
Outline Give only the key facts, or the steps of a process in the correct order.
Predict Look at the data and suggest a sensible value or outcome. Use trends in the data and your science knowledge to help you.
Show Write down the details, steps, or calculations to show how to get an answer.
Suggest Apply something you have learnt to a new context, or to come up with a reasonable answer.
Write down Give a short answer. There is no need for an argument to support your answer.

Answer	Examiners' comments
Exersise is good for you it makes you feel better and you can run faster. It can help you loose wait.	**Grade G** answer: this answer makes some valid points but it does not address all of the question as it does not mention the effect of diet. There are mistakes in spelling, punctuation, and grammar.
Diet supplies the energy for movement and growth. It also supplies the nutrients that are needed to build the muscles and bones. It is important to have enough of the right nutrients to enable the body to grow properly and to be able to carry out all the active processes a healthy person needs to do. However, it is also important not to have too much of any one nutrient as this can lead to problems such as obesity. *Exercise is also important as it uses the energy that we eat; if we eat more, more exercise can use up the extra energy. If we gain weight we can (and should) lose the excess weight by eating less and exercising more - the exercise will use up energy that has been stored as fat. However, if we are at a healthy weight and exercise a lot we must eat more to ensure that we do not lose too much weight.*	**Grade A/A*** answer: the arguments are made clearly and are organised logically. The candidate has discussed diet first then moved on to the effects of exercise. The candidate has also covered a range of scenarios where diet and exercise do not balance. The spelling, punctuation, and grammar are faultless.

Equations, units, and data

Equations

You might need to use these equations in the exam. They will be on the exam paper, so you do not need to learn them off by heart.

B4 The processes of life

Photosynthesis

$$6CO_2 + 6H_2O \xrightarrow[\text{light energy}]{} C_6H_{12}O_6 + 6O_2$$

Aerobic respiration

$$C_6H_{12}O_6 + 6O_2 \longrightarrow 6CO_2 + 6H_2O$$

B7 Further biology

$$\text{Body mass index (BMI)} = \frac{\text{body mass (kg)}}{[\text{height (m)}]^2}$$

Units

Length: metres (m), kilometres (km), centimetres (cm), millimetres (mm), micrometres (μm), nanometres (nm)

Mass: kilograms (kg), grams (g), milligrams (mg)

Time: seconds (s), milliseconds (ms), hours (h)

Temperature: degrees Celsius (°C)

Area: cm^2, m^2

Volume: cm^3, dm^3, m^3, litres (l), millilitres (ml)

Energy: joules (J), kilojoules (kJ), megajoules (MJ), kilowatt-hours (kWh), megawatt-hours (MWh)

Power: watts (W), kilowatts (kW), megawatts (MW)

B1 You and your genes workout

1 Use these words to finish labelling the diagram.

cell **nucleus** **chromosomes** **genes** **DNA**

a ____________________

b ____________________

c ____________________

d ____________________

made of long

e ____________________ molecules

2 Draw lines to match each characteristic to the factors that determine it.

Characteristics
dimples
weight
eye colour
scars

Factors
genes and environment
one gene only
environment only
several genes working together

Exam tip

Make sure you know what genes, chromosomes, and DNA are.

3 Write the letter **T** next to the statements that are true.
Write the letter **F** next to the statements that are false.

a Women have two X chromosomes in each cell, except for their sex cells. ______

b Men have one Y chromosome in each cell. ______

c Human egg cells contain 46 chromosomes. ______

d Every sperm has an X chromosome. Half of all sperm also have a Y chromosome. ______

e If a sperm with a Y chromosome fertilises an egg, the embryo develops female sex organs. ______

4 Choose words from the box to fill in the gaps.

clones	genes	unspecialised	environments	asexual	sexual

Some strawberry plant cells are ____________. These cells can grow new plants. This is what happens in ____________ reproduction. The new strawberry plants have genes that are exactly the same as their parent's genes. They are ____________ of the parent plant. In this case, all the variation between the strawberry plant and its offspring are caused by differences in their ____________.

5 For each use of genetic testing listed in the table, write down one question the person having the test might need to think about. The first one has been done for you.

Use of genetic testing	Question to think about
For a man to find out if he will develop symptoms of a genetic disorder such as Huntington's disease.	Should I tell my employer if the test result is positive?
a For a pregnant woman to find out if her fetus has two faulty alleles for cystic fibrosis.	
b For a man to find out if he is a carrier of a genetic disorder.	

6 The allele that gives you hair on the middle of your fingers is dominant (R). The allele for no hair is recessive (r).

a Complete the genetic diagram (Punnett square) for the mother and father shown.

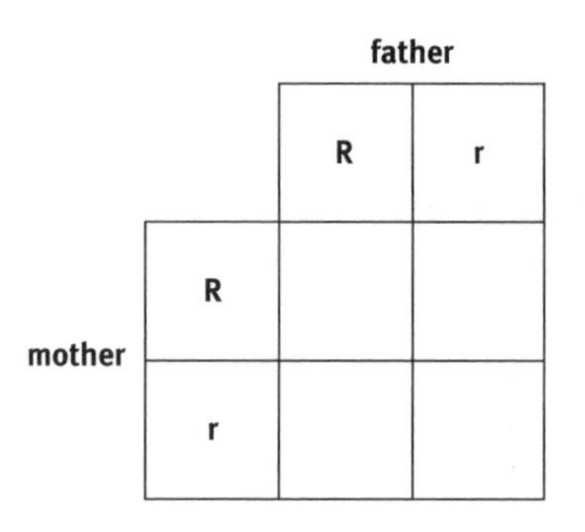

b Use the genetic diagram to help you complete the sentences below.

i The chance of a child of the mother and father having hairy fingers is ________ %.

ii The chance of a child of the mother and father not having hairy fingers is ________ %.

7 Draw a line to link two words on the circle.
Write a sentence on the line saying how the two words are connected.
Repeat for as many words as you can.

chromosome

gene

protein

environment

allele

dominant

recessive

DNA

clones

asexual

sexual

stem cell

B1.1.1–6 What are genes?

Every living organism is made from **cells**. Most cells have a **nucleus**. Inside the nucleus are **chromosomes**. Chromosomes are made from very long molecules of DNA.

Chromosomes contain thousands of **genes**. Genes are instructions that control how a living thing develops and functions. Genes control which proteins a cell makes, including:

- **structural** proteins to build the body, for example collagen
- **functional** proteins to take part in chemical reactions, for example enzymes such as amylase.

Your characteristics depend on your genes and the **environment**.

- Some depend on one gene only, for example dimples.
- Some depend on several genes working together, for example eye colour.
- Some depend on the environment only, for example scars.
- Some depend on genes and the environment, for example weight.

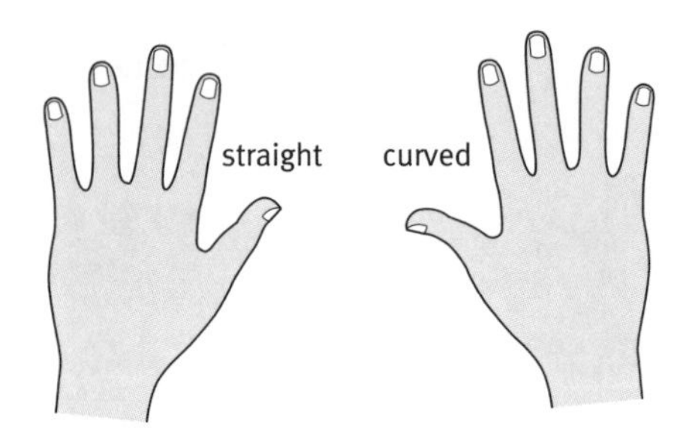

The allele that gives you straight thumbs is dominant. The allele that gives you curved thumbs is recessive.

B1.2.1–6 Why do family members look alike?

Human cells (except sex cells) contain pairs of chromosomes. One chromosome in each pair came from the mother's egg and the other came from the father's sperm.

In the two chromosomes of a pair, the same genes are in the same place. The two genes in a pair can be different. Different versions of the same gene are called **alleles**. For each gene, a person has either two identical alleles or two different alleles.

If both alleles for a gene are the same, you are **homozygous** for that gene. If both alleles are different, you are **heterozygous**.

You have similarities to your parents because you developed from a fertilised egg that got alleles from your mother and your father.

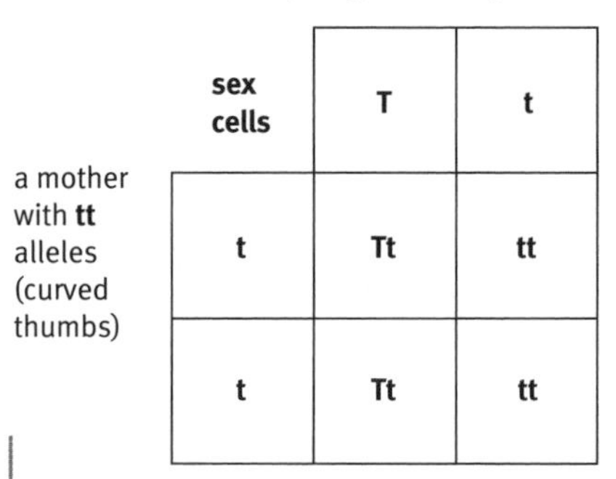

a father with **Tt** alleles (straight thumbs)

a mother with **tt** alleles (curved thumbs)

sex cells	**T**	**t**
t	**Tt**	**tt**
t	**Tt**	**tt**

There is a 50% chance that a child of these parents will have straight thumbs.

B1.2.7–8, B1.2.12 What makes us different?

Brothers and sisters are not identical. This is because they inherit different combinations of alleles from their parents.

Alleles can be **dominant** or **recessive**. For example, the allele that gives you straight thumbs (**T**) is dominant. The curved thumb allele (**t**) is recessive. If you inherit either one or two dominant alleles (**Tt** or **TT**) you will have straight thumbs. If you inherit two recessive alleles (**tt**) you will have curved thumbs.

A description of your genes, including your combination of alleles, is called your **genotype**. Your **phenotype** is a description of your characteristics, for example whether or not your thumbs are curved.

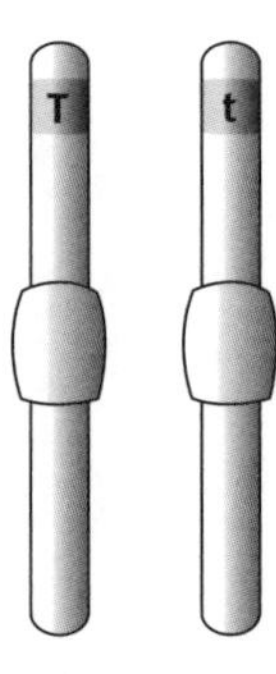

These chromosomes are a pair. The gene controlling thumb shape is shaded. The person inherited one **T** and one **t** allele. He has straight thumbs.

B1.2.9–10 What makes us male or female?

Humans have one pair of sex chromosomes in each body cell. Female humans have two X chromosomes (XX). Males have one X chromosome and one Y chromosome (XY).

The chromosomes in sex cells are not paired up. Every egg has an X chromosome. Half of all sperm have an X chromosome. The other sperm have a Y chromosome.

H The Y chromosome includes a sex-determining gene. This makes an embryo develop testes, and so become male. When there is no Y chromosome, the embryo develops ovaries. It is female.

Exam tip

Practise drawing and interpreting genetic diagrams (Punnett squares).

B 1

B1.3.1–5 What causes inherited disorders?

Some disorders are caused by inheriting faulty alleles of just one gene.

Huntington's disease develops after age 35. It is fatal.

Symptoms:
- tremor and clumsiness
- memory loss and concentration problems
- mood changes

One faulty dominant allele – **H** – causes Huntington's disease. A person can inherit the disease from just one parent.

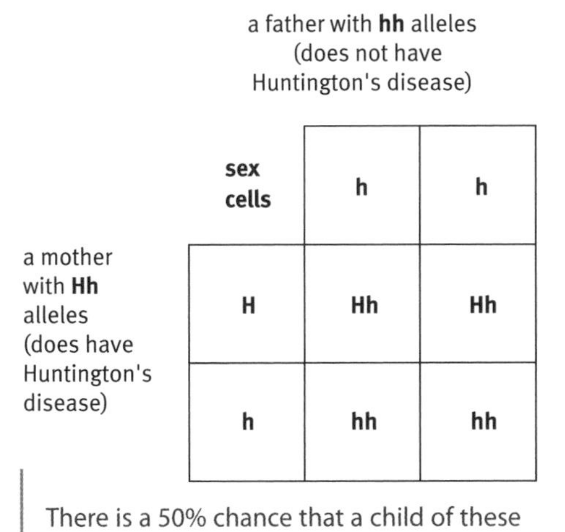

There is a 50% chance that a child of these parents will inherit Huntington's disease.

Children with **cystic fibrosis** make thick, sticky mucus. They have:
- difficulty breathing
- difficulty digesting food
- frequent chest infections.

A faulty recessive allele – **f** – causes cystic fibrosis. A child who has two faulty alleles – one from each parent – has cystic fibrosis. A child with one faulty allele is a **carrier**.

Carriers do not have the disorder, but can pass it on to their children.

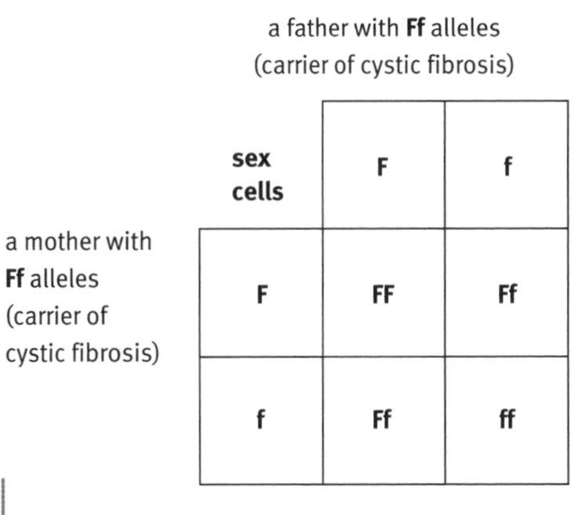

There is a 25% chance that a child of these parents will inherit cystic fibrosis. There is a 50% chance that a child will be a carrier of cystic fibrosis.

B1.3.6–8 Why have genetic tests?

Genetic tests detect faulty alleles.

Reason for test	Issues to consider
For an adult to find out if they are a carrier of a genetic disorder.	Should a carrier have children?
For an adult to find out if they will develop symptoms of a genetic disorder.	Should the adult tell relatives who may also develop symptoms?
For a pregnant woman to find out if her fetus has a genetic disorder.	Should she have an abortion if the test is positive? Will removing cells for the test increase the risk of miscarriage?
To find out how a person might react to a prescription drug.	Should a different drug be used? Does the person need a higher dose?
H To find out if embryos created outside the body have faulty alleles, and so choose which to implant in the uterus. This is pre-implantation genetic diagnosis.	Does removing a cell to test damage an embryo? Is it ethically right to destroy embryos with faulty alleles?

The results from genetic tests can be inaccurate. **False positives** wrongly show that a person will develop a disorder. **False negatives** wrongly show that a person will not develop a disorder.

B1.4.1–4 What are clones?

Bacteria, plants, and some animals can reproduce asexually to form clones. **Clones** are individuals with identical genes. Only environmental factors can cause differences between clones.

Plants that produce bulbs or runners are clones of each other. Animals do not usually form clones, but there are exceptions:

- Identical twins are clones. They form when the cells of an embryo separate to make two embryos.
- H Scientists make clones by removing an egg cell nucleus. They take a nucleus from an adult body cell of the organism they want to clone, and transfer it to the 'empty' egg cell. They grow the embryo for a few days and implant it into a uterus.

B1.4.5–7 What are stem cells?

Stem cells are unspecialised cells. There are two types.

- **Embryonic stem cells** can develop into any type of cell.
- **Adult stem cells** can develop into many, but not all, types of cell.

Most stem cells become specialised during the early development of a living organism.

Because they are unspecialised, stem cells may be useful in future to treat some illnesses.

Use extra paper to answer these questions if you need to.

1 Choose words from the box to fill in the gaps. Each word may be used once, or not at all.

alleles	**instructions**	**proteins**	**characteristic**
carbohydrates	**chromosomes**	**genes**	**DNA**

Living things are made from cells. Inside every cell nucleus are long threads called __________. These contain thousands of __________. Genes are __________ that control how a living thing will develop. They are codes for making __________. Genes are sections of very long __________ molecules.

2 Highlight the statements below that are **true**. Then write corrected versions of the statements that are **false**.

a Body cells contain pairs of chromosomes.

b Different versions of the same gene are called alleles.

c If a person has one dominant allele in a pair of alleles, they will not show the characteristic linked to that gene.

d Human male body cells have XX sex chromosomes.

e A sperm contains pairs of chromosomes.

3 Circle four symptoms of cystic fibrosis from those below.

thick mucus	**tremor**
clumsiness	**difficulty breathing**
chest infections	**memory loss**
digestion problems	**blood in urine**

4 Jess is a carrier of cystic fibrosis. She is pregnant. She is trying to decide whether or not to have a genetic test to find out if her fetus has cystic fibrosis. List three things she might want to consider before having the test.

5 Marfan syndrome is an inherited disorder. It may cause serious heart and eyesight problems. It is caused by faulty alleles (**M**) of a single gene. The faulty allele is dominant.

a Which of the people listed below will have Marfan syndrome?
Ellie (**Mm**), Jim (**mm**), Susan (**MM**), Tom (**mM**)

b Ellie and Jim are having a baby. Draw a Punnett square to work out the chance of the baby having Marfan syndrome.

6 Assume that being able to roll your tongue – or not – is controlled by one pair of alleles. The ability to roll your tongue is dominant (**R**). Not being able to roll your tongue is recessive (**r**).

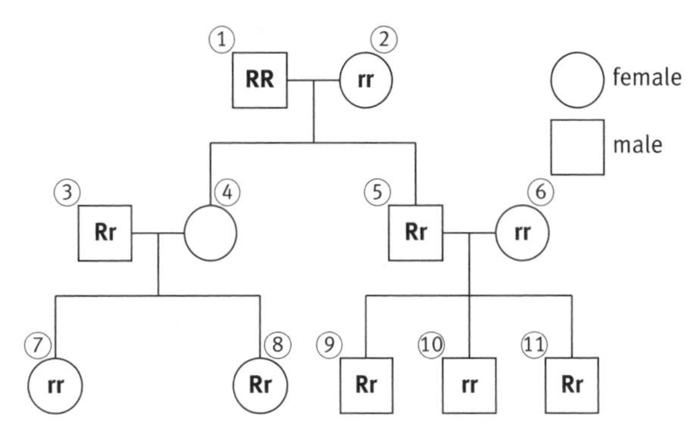

On the family tree above, each person is represented by a number.

a Write down the pair of alleles that person 4 has.

b Write the numbers of the people who cannot roll their tongues.

c Write the numbers of the people who can roll their tongues themselves, but who might have children who cannot roll their tongues.

7 List five symptoms of Huntington's disease.

8 What is a stem cell? Explain the difference between adult and embryonic stem cells.

H 9 Look at the Punnett square below.

a Use it to help you describe the difference between a person's genotype and their phenotype.

b Explain how the Punnett square shows that the chance of a baby being female is 50%.

		mother	
		X	X
father	X	XX	XX
	Y	XY	XY

10 Natasha has dimples. Her genotype for dimples is **Dd**.

a What is Natasha's phenotype for dimples?

b Is the dimple allele dominant or recessive? Give a reason for your decision.

c Is Natasha homozygous or heterozygous for dimples? Explain how you know.

11 Explain how the sex-determining gene on the Y chromosome makes an embryo develop into a male.

12 a Explain what is meant by *pre-implantation genetic diagnosis.*

b Suggest why a man and woman who want a baby may choose to go through this process.

c Identify the issues the man and woman might need to consider before going through the process.

13 Describe the stages involved in creating an artificial animal clone.

1 Ellen and Hannah are identical twin girls.

a **i** Ellen and Hannah look the same as each other.
Choose the best explanations for this.
Put ticks in the **two** correct boxes.

They have the same combination of alleles. ☐

They inherited genes from both parents. ☐

They both developed from one egg that was fertilised by one sperm. ☐

Their parents have different combinationsof alleles. ☐ [1]

ii Ellen and Hannah look different from their mother.
Choose the **wrong** explanation for this.
Put a tick in **one** box below.

A person's characteristics are affected by both genes and the environment. ☐

They received alleles from both parents. ☐

The twins and their mother have different combinations of alleles. ☐

Their cells contain 23 pairs of chromosomes. ☐ [1]

iii Ellen has one pair of sex chromosomes in each body cell.
Which two chromosomes are in this pair?
Circle the correct answer.

XY **YY** **XX** [1]

b **i** John and Jim are identical twins. They are 50 years old.
John is fatter than Jim. Choose the best explanation for this.
Put a tick in the **one** correct box.

They have different combinations of alleles. ☐

They are clones of each other. ☐

They have different lifestyles. ☐

John was born an hour before Jim. ☐ [1]

ii John and Jim have stem cells in many of their body tissues.
What is a stem cell? [1]

iii Suggest one way in which stem cells might be used in future. [1]

Total [6]

2 The allele that causes straight thumbs is dominant (**T**).
The allele that causes curved thumbs is recessive (t).

Sarah has straight thumbs. She has one **T** allele and one **t** allele.
Alan has curved thumbs. He has **tt** alleles.

a **i** What percentage of Sarah's egg cells contain the allele **T**?

______________________________ [1]

ii Give the number of **t** alleles in each of Alan's body cells (*not* the number in his sperm cells).

______________________________ [1]

b **i** Finish the punnett square to show which alleles Sarah and Alan's children may inherit. [3]

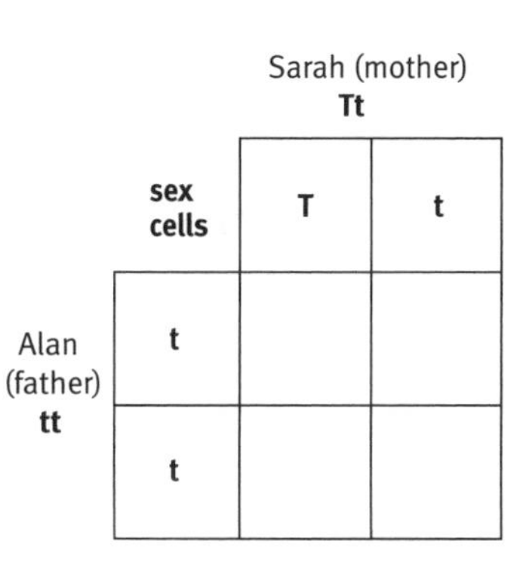

ii Sarah and Alan have a baby boy.
What is the chance of his having a straight thumb?
Put a ring round the correct answer.

25% **50%** **75%** **100%** [1]

Total [6]

3 Huntington's disease is an inherited disease.
Its symptoms usually develop after the age of 35.

a Give two symptoms of Huntington's disease.

______________________________ [2]

b Huntington's disease is caused by a dominant allele of just one gene. The table shows the alleles of this gene in the cells of four people.

Name	Alleles
Abigail	**Hh**
Brenda	**HH**
Chris	**hh**
Deepa	**hh**

Who will develop Huntington's disease?
Circle the correct name or names.

Abigail **Brenda** **Chris** **Deepa** [1]

c Gary is 20. He had a genetic test. The test shows that he will develop Huntington's disease.

Gary's wife is six weeks pregnant. Suggest one reason why the couple may decide to test the fetus for Huntington's disease. Suggest one reason why they may decide not to have this test.

For the test: ______________________________

____

Against the test: ______________________________

______________________________ [2]

Total [5]

4 Use ideas about genes and alleles, and body cells and sex cells, to explain why member of a family of two parents and their two children show variation.

✎ The quality of written communication will be assessed in your answer to this question.

Write your answer on separate paper or in your exercise book.

Total [6]

5 Niemann-Pick disease is an inherited disorder. It is caused by a faulty allele. The faulty gene causes reduced appetite, unsteady walking, and slurring of speech.

Look at the family tree.

Rod
Barbara
Sally
Philip
Tracy
Andrew
Kylie
Mitch
Olivia

key
female without Niemann-Pick
female with Niemann-Pick
male without Niemann-Pick
male with Niemann-Pick

a Explain how the family tree shows that the faulty allele is recessive. [2]

b Name the people who are definitely carriers of the disease. [1]

Going for the highest grades

c Name the people who must be homozygous for Niemann-Pick disease. [1]

d Use the example of Niemann-Pick disorder to help you explain the difference between the words *phenotype* and *genotype*. [2]

e Niemann-Pick is a serious disorder. It can cause death in childhood.

Sally and Matt want a baby. They are carriers of Niemann-Pick disorder. Their doctor suggests using their sperm and eggs to create embryos outside the body, and then performing pre-implantation genetic diagnosis on the embryos.

i Explain the meaning of the term *pre-implantation genetic diagnosis*. [2]

ii Describe some implications of pre-implantation genetic diagnosis. [3]

Write your answer on a separate piece of paper or in your exercise book.

Total [11]

Exam tip

Practise interpreting family tree diagrams.

B2 Keeping healthy

workout

1 Write each phrase from the box in a sensible place on the flow diagram.

damage cells
disease symptoms
reproduce rapidly
make toxins

Harmful microorganisms enter the body.

or

2 Here are the stages in making an influenza (flu) vaccine.

- Experts meet every April to decide which strain of wild flu virus is likely to attack next winter.
- In labs, scientists make a special 'hybrid' flu virus.
- This flu virus is delivered around the world.
- Technicians drill holes in fertilised hens' eggs.
- Technicians inject the flu virus into the eggs and seal the hole with wax.
- The eggs provide food and moisture. They are kept warm at about 37 °C for 10 days.
- Technicians harvest the flu virus from the eggs.
- Technicians break the flu virus into pieces and put it into the vaccine.

a Underline the stage that takes account of the fact that the flu virus changes very quickly.

b Draw a box around the stage that shows the conditions the flu virus needs to reproduce quickly.

c Draw a cloud around the stage that shows how the virus is made safe before being put into the vaccine.

d Draw a triangle around the stage that indicates that the flu virus spreads easily from person to person.

B2

3 Look at captions **A** to **L** below. Write one letter in each box to show how water balance is controlled in the body. Two have been done for you.

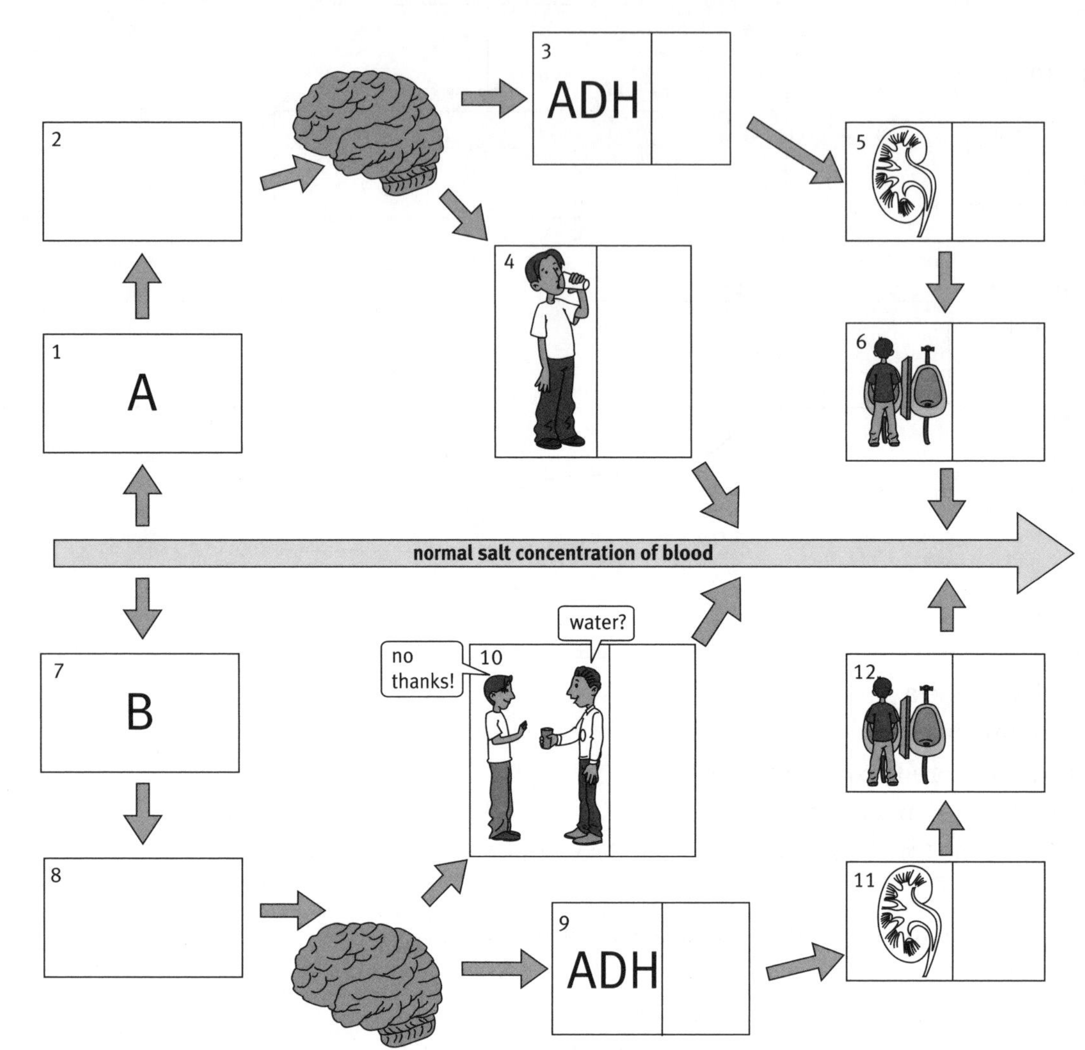

A The concentration of the blood plasma increases.

B The concentration of the blood plasma decreases.

C Receptors in the brain are stimulated.

D ADH is secreted by the pituitary gland.

E More water is reabsorbed in the kidneys.

F Less urine is made. It is concentrated.

G Receptors in the brain are not stimulated.

H Less ADH is secreted by the pituitary gland.

I Less water is reabsorbed by the kidneys.

J More urine is made. It is dilute.

K The person feels thirsty and drinks water.

L The person does not feel thirsty, so drinks little water.

B2.1.1–6 How do our bodies resist infection?

Inside your body, conditions are ideal for **microorganisms** like bacteria and viruses. So they reproduce quickly. Some bacteria and viruses cause infectious diseases. They give you disease symptoms by damaging cells or making poisons (**toxins**).

White blood cells try to destroy harmful microorganisms. They are part of your **immune system**. One type of white blood cell **engulfs** and **digests** harmful microorganisms. Another type of white blood cell makes **antibodies**.

Some of the white blood cells that make each antibody stay in your blood. These are **memory cells**. If the same microorganism invades your body in future, memory cells recognise its antigens. The memory cells quickly make the correct antibodies. The invaders are destroyed before you feel ill. You are **immune** to the disease.

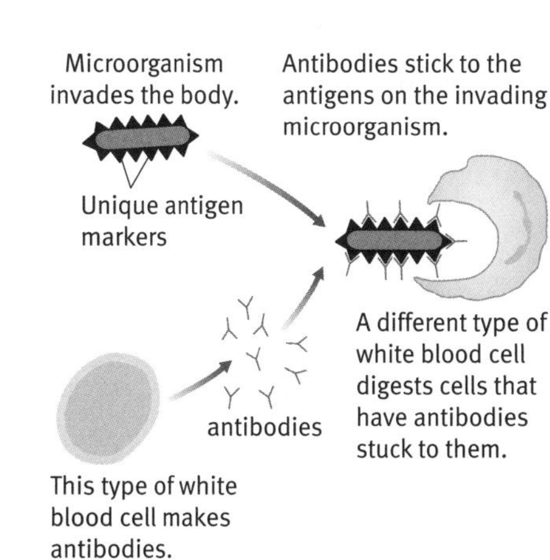

One type of white blood cell makes antibodies to label microorganisms. A different type digests the microorganisms.

B2.2.1–5 How do vaccines work?

Vaccines prevent you getting diseases. A vaccine contains a safe form of a disease-causing microorganism. When a vaccine is injected into your body, white blood cells make antibodies against the microorganism. Your body stores some of these memory cells. If an active form of the microorganism enters your blood in future, memory cells quickly make antibodies. You do not get ill.

Having vaccinations is not risk free. Different people have different side-effects from vaccinations because they have different genes.

H To prevent epidemics of infectious diseases, a high percentage of the population must be vaccinated. If they are not, large numbers of the disease-causing microorganism will remain in infected people, and people who cannot be vaccinated are likely to catch the disease.

B2.2.6–10 How do antibiotics work?

Antimicrobials are chemicals that may kill bacteria, fungi, and viruses.

H Some types of antimicrobial **inhibit** the reproduction of microorganisms.

Antibiotics are effective against bacteria, but not viruses.

Over time, some bacteria and fungi become **resistant** to antimicrobials. To reduce antibiotic resistance, people must:

- finish all the tablets, even if they feel better
- only use antibiotics when necessary.

Antimicrobial resistance develops when random changes (mutations) in the genes of bacteria or fungi make new varieties that the antimicrobial cannot kill or inhibit.

The bacterial population is not resistant to the antibiotic.

Mutations in some bacteria make them resistant to the antibiotic. There is variation in the population.

Part-way through the course most of the bacteria are killed. Only those with most resistance are still alive.

The person feels better, so they stop taking their antibiotics.

With no competition for space and food from other bacteria, resistant ones can reproduce quickly.

The whole bacteria population is now resistant to the antibiotic.

A few mutations can result in antibiotic-resistant bacteria.

B2.2.11–15 How are new drugs trialled?

In most human trials on ill people, one group of patients takes the new drug. Another group of patients are **controls**. The controls take either the existing treatment for the illness, or a placebo.

A **placebo** looks like the new treatment, but has no drugs in it. Placebos are rarely used in human trials because people who take them miss out on the benefits of both new and existing treatments.

There are three types of human trial:

- In **double-blind** trials, neither patients nor doctors know who is in which group.
- In **blind** trials, doctors know who is in which group, but patients do not.
- In an **open-label** trial both the patient and their doctor knows whether the patient is given the new drug. These trials are used when the patient cannot affect the outcome of using the drug.

Long-term human trials ensure that the drug is safe and that it works. They also identify side-effects that do not occur immediately.

A new drug is tested for safety and effectiveness on lab-grown human cells and animals.

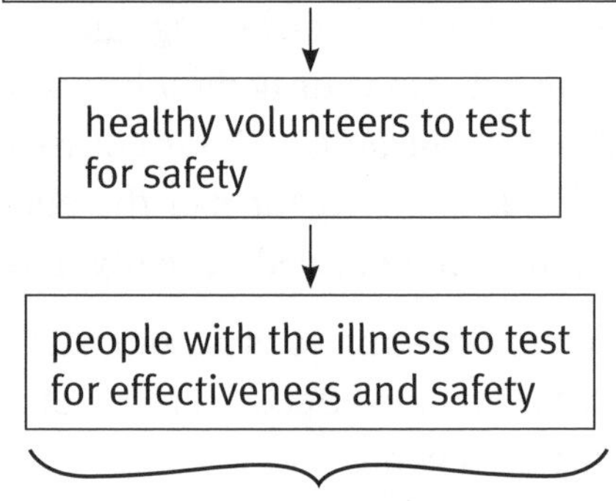

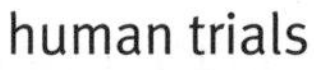

B2.3.1–7 How does your heart work?

Your heart pumps blood around your body. It is a **double pump**:

- The right lower chamber pumps blood to your lungs.
- The left lower chamber pumps blood to the rest of your body.

Your heart is made from muscle. It has its own blood supply. The blood brings oxygen and glucose to the heart. Heart cells use these as a supply of energy.

Blood travels around your body through **arteries**, **veins**, and **capillaries**. Blood vessels are well adapted to their functions:

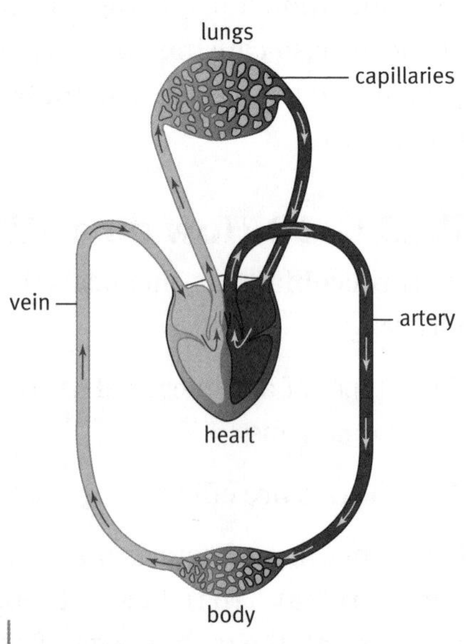

Blood flow around the body.

- **Arteries** take blood from the heart to your body. Their walls withstand the high pressure created by the pumping heart.

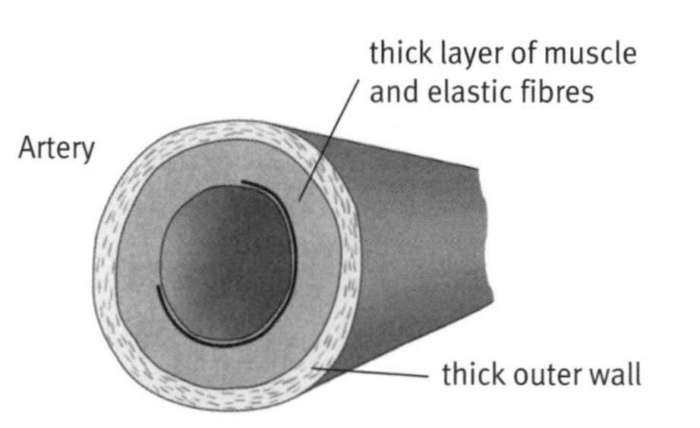

- **Veins** have thin walls that can be squashed when you move. This pushes blood back to the heart.

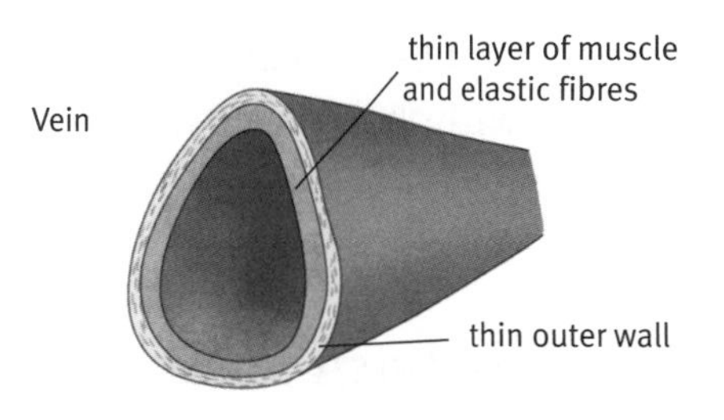

- **Capillaries** take blood to and from tissues. Their very thin walls allow oxygen and food to diffuse into cells and waste to diffuse out of cells.

thin wall (one cell thick)

5–20 μm diameter

Exam tip

Revise the differences between arteries, veins, and capillaries carefully.

Doctors can check how well your heart is working. They measure:

- **pulse rate** at your wrist (the number of beats per minute)
- **blood pressure** – this measures the pressure of the blood on the walls of your arteries. It has two numbers. The higher number is the pressure when the heart is contracting. The lower number is the pressure when the heart is relaxing. Height, weight, gender, and lifestyle affect blood pressure, so there is a range of 'normal' blood pressure.

B2.3.8–16 What causes heart attacks?

Coronary arteries carry oxygenated blood to the heart. Fat can build up on the artery walls. A blood clot may form on this fat. This may block the artery. The blockage stops oxygen getting to the heart muscle. Heart cells die, and the heart is permanently damaged. This is a **heart attack**.

Your genes and your **lifestyle** affect your chances of having a heart attack. Poor diet, stress, and high blood pressure increase the risk of heart disease. So does misusing drugs such as nicotine (in cigarettes), alcohol, cannabis, and Ecstasy. Regular exercise reduces the risk of heart disease.

Heart disease is more common in the UK than in less industrialised countries. This may be because British people eat less healthily, or have higher levels of stress or drug misuse.

Scientists use **epidemiological studies** to identify risk factors for heart disease. These studies look at large numbers of people.

They often compare groups of people, for example, smokers and non-smokers. The two groups of people must be **matched**. This means they should be similar apart from the factor being tested.

In large-scale **genetic** studies, scientists compare the genes of people with a disease with those of healthy people. This helps identify alleles that increase the risk of a person getting the disease.

B2.4.1–5 What is homeostasis?

Automatic control systems keep body conditions constant. They are controlled by nervous and hormonal communication systems. Keeping a constant internal environment is called **homeostasis**.

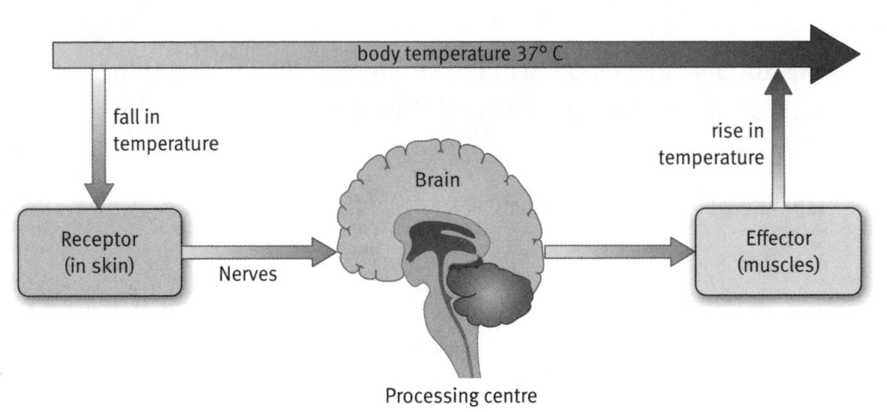

The diagram shows how body temperature is controlled automatically.

Body control systems have:

- **receptors** to detect changes in the environment
- **processing centres** to receive information and coordinate responses automatically
- **effectors** to produce responses.

Your body's temperature and water control systems are automatic.

H They use **negative feedback** between the effector and the receptor of a control system to reverse changes to the system's steady state.

B2.4.6–13 How do we control water levels?

Cells only work properly if the concentrations of their contents are correct. So their water levels must be kept constant.

Your kidneys get rid of waste products by **excretion**. They also control water levels in your body. They do this by responding to water levels in **blood plasma**.

Water levels in your blood plasma may go down because of:

- sweating after exercise or on a hot day
- eating salty food
- not drinking much water.

When water levels in blood plasma are low, your kidneys make less urine. The urine is concentrated, so it is dark. When water levels in blood plasma are high, your kidneys make lots of dilute urine.

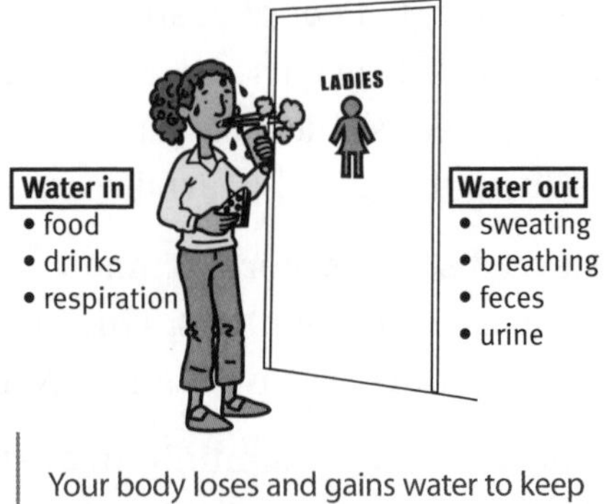

Your body loses and gains water to keep water levels balanced.

Kidneys are part of a negative feedback system.

- Receptors in the brain detect changes in concentration in blood plasma.
- If the concentration is too high, the **pituitary gland** in the brain releases **ADH** (a hormone) into the blood stream.
- The ADH travels to the kidneys (**effectors**). The more ADH that arrives, the more water the kidneys reabsorb into the body. So the more concentrated the urine.

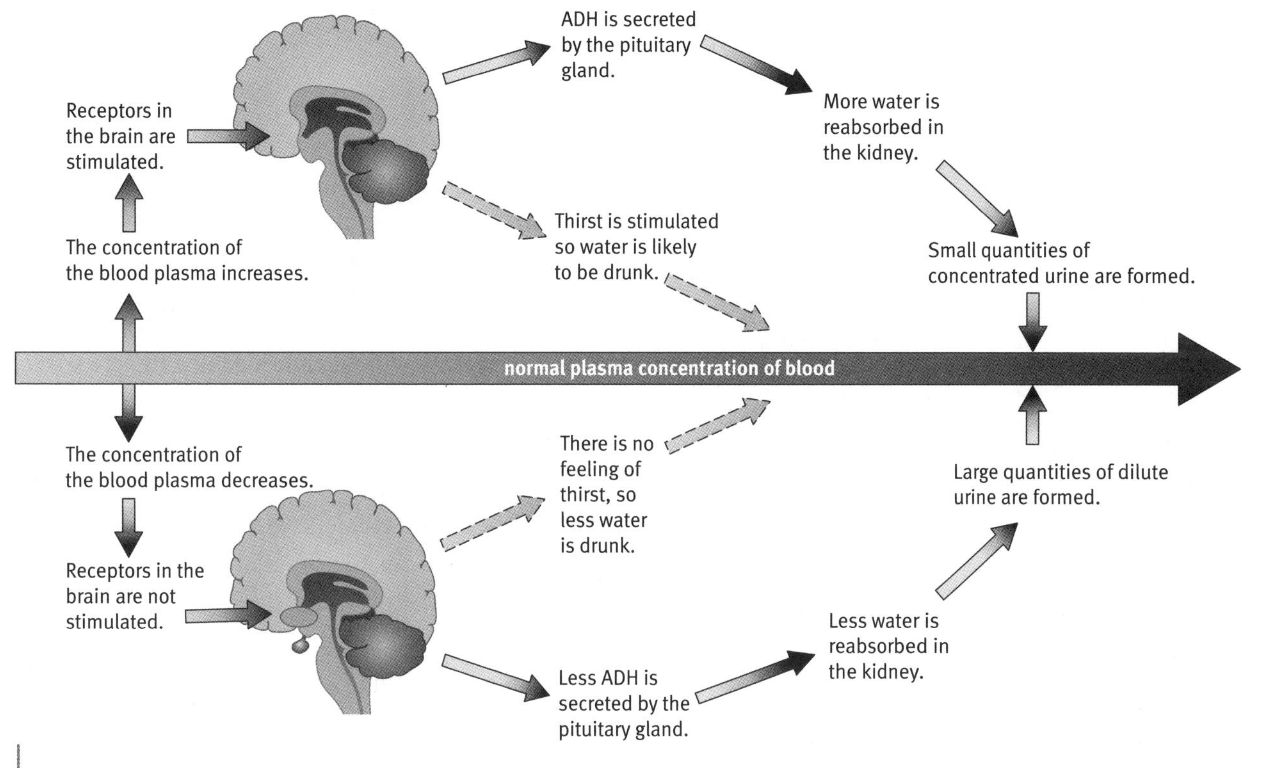

Water balance is controlled by negative feedback, involving both nervous and hormonal communication.

Drugs affect the amount of urine you make:

- Alcohol leads to big volumes of dilute urine, so you may get dehydrated.
  This is because alcohol stops the pituitary gland releasing ADH.
- Taking Ecstasy leads to small volumes of concentrated urine.
  This is because Ecstasy makes the pituitary gland release more ADH.

Use extra paper to answer these questions if you need to.

1 Tick the boxes to show how to reduce the risk of heart disease.

a Eat more fatty food. ☐
b Do not smoke. ☐
c Avoid stress. ☐
d Do not exercise. ☐
e Smoke cannabis occasionally. ☐
f Keep your blood pressure high. ☐

2 Write an L next to the processes by which your body loses water, and a G next to the processes by which your body gains water.

a eating food **c** urinating **e** producing feces
b sweating **d** breathing **f** respiration

3 Draw lines to match each measurement with its definition.

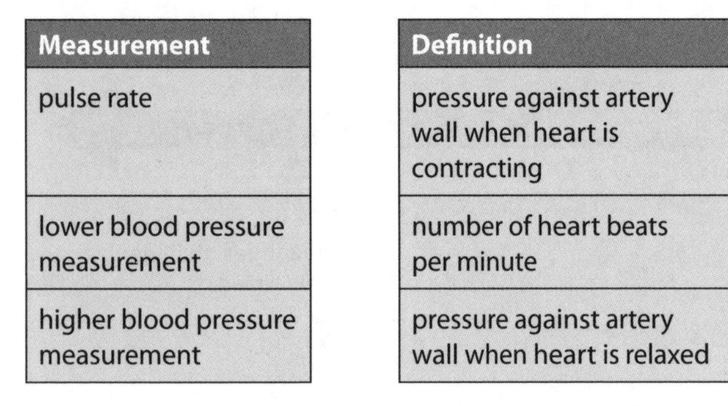

Measurement	Definition
pulse rate	pressure against artery wall when heart is contracting
lower blood pressure measurement	number of heart beats per minute
higher blood pressure measurement	pressure against artery wall when heart is relaxed

4 Highlight the statements below that are **true**. Then write corrected versions of the statements that are **false**.

a Antibiotics kill viruses.
b Antimicrobials kill viruses, bacteria, and fungi.
c New drugs are tested for effectiveness on human cells grown in the lab.
d In clinical trials, new drugs are tested for effectiveness on healthy volunteers.
e A placebo has no drugs in it.

5 Complete the table.

Part of circulation system	What does it do?	What is it made from?
heart	pumps blood around the body	
artery		
vein		thin walls made of muscle and elastic fibres
capillary		

6 The steps below describe how a vaccine works. Write the letters of the steps in the best order.

A The vaccine is made from a safe form of the virus.
B White blood cells digest the clump.
C The vaccine is injected into the person.
D Sometime later an active form of the virus gets into the blood.
E Memory cells quickly make the correct antibodies.
F White blood cells make antibodies that stick to the antigens on the safe form of the virus. Some of these white blood cells are stored in the body as memory cells.
G The antibodies make the viruses clump together.

7 Draw lines to match each part of the control system to what it does.

Part of control system	What it does
receptor	receives information and processes responses
processing centre	detects changes in environment
effector	produces the response

8 Explain why taking vaccines and medicines is not risk free.

9 List two things people can do to help to reduce antibiotic resistance.

H **10** Write a P next to any trial types in which the patient knows whether or not she is in the control group. Write a D next to any trial types in which the doctor knows whether or not the patient is in the control group.

a double-blind trial
b blind trial
c open-label trial

11 Explain the purpose of long-term human trials for new drugs.

12 MRSA is a type of bacteria that is resistant to most types of antibiotics. Explain how resistant bacteria develop.

13 The diagram shows the negative feedback system in a premature babies' incubator. Annotate the diagram to show how it is similar to the negative feedback system that controls water levels in the human body.

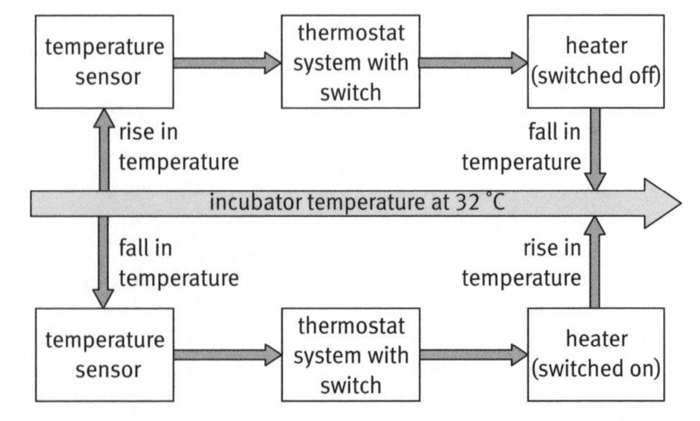

14 Explain the effects that alcohol and Ecstasy have on levels of ADH, and how this affects urine production.

1 Lauren has food poisoning. She has diarrhoea and vomits frequently.

She became ill after she ate a raw egg that contained *Salmonella* bacteria.

a The graph shows the changes in the number of *Salmonella* bacteria in Lauren's stomach.

number of *Salmonella* bacteria in Lauren's stomach (millions)

0 1 2 3 4 5

1 2 3 4 5 6 7

time (days)

i Name the process that causes the number of bacteria to increase during the first few hours.

____ [1]

ii Use the graph to complete the sentence below.

Lauren will probably begin to feel better ______ days after she ate the raw egg.

b Lauren's body tries to get rid of *Salmonella* bacteria in two ways:

- Vomiting and diarrhoea remove some of the bacteria from the intestines.
- Certain blood cells can destroy the bacteria.

i Name the type of blood cells that can destroy *Salmonella* bacteria.

____ [1]

ii Suggest why Lauren's doctor advised her **not** to take anti-diarrhoea tablets.

____

____ [1]

Total [3]

2 The graph shows the percentage of British 2-year-olds who received the MMR vaccine from 1989 to 2002. The MMR vaccine prevents people getting measles, mumps, and rubella.

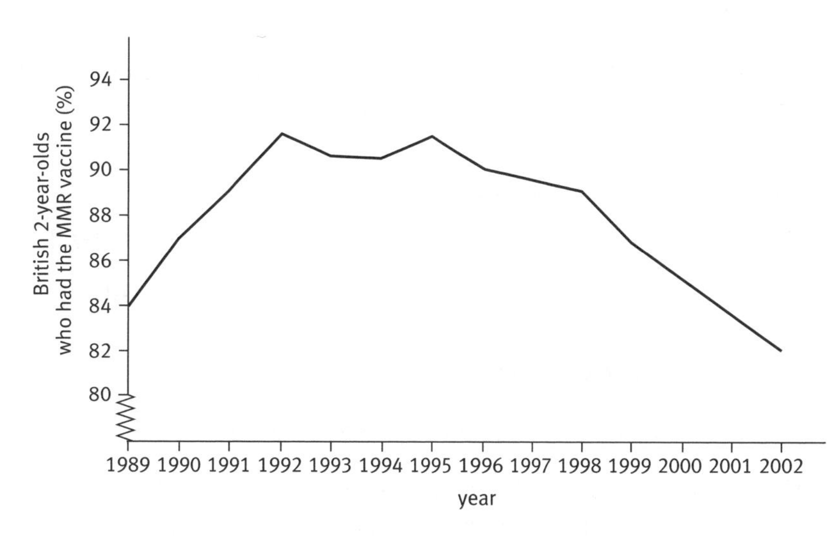

a Here are some people's opinions about the triple MMR vaccine.

A I'm worried about the vaccine's possible serious side effects on my child.

B I'm a doctor. No vaccine is completely safe. The side effects of the MMR vaccine are a possible risk, but the dangers of measles, mumps, and rubella are worse.

C Measles is a nasty disease. I don't want to risk my child getting it.

D The more children who have the MMR vaccine, the better. Then everyone is protected from measles, mumps, and rubella.

Use the graph and the opinions **A, B, C, and D** to complete the sentences below.

Between 1989 and 1992 the percentage of children who had the MMR vaccine ________________ .

One opinion that may explain this trend is opinion ________________ . Between 1992 and 2001 the percentage of children who had the MMR vaccine ________________ . One opinion that may explain this trend is opinion ________________ . [2]

b Matthew had the MMR vaccine when he was one year old. Two years later, the measles virus got into his body.

Matthew did not get measles. The stages below explain why.

A A nurse injects Matthew with the MMR vaccine. The vaccine contains safe forms of measles, mumps, and rubella viruses.

B These cells make antibodies very quickly.

C The natural measles virus gets into Matthew's bloodstream.

D The virus is destroyed before it has time to make Matthew feel ill.

E Matthew's white blood cells make antibodies to recognise measles, mumps, and rubella viruses.

F Memory cells recognise the virus.

G The level of antibodies in Matthew's blood falls over the next two years.

The stages are in the wrong order. Write a letter in each empty box to show the correct order.

A					

[4]

Total [6]

Answer this question if you expect to take the Foundation tier test.

3 Explain why and how water levels in the cells of a human body are kept constant, and how alcohol and Ecstasy affect the amount and concentration of urine a person makes.

✎ The quality of written communication will be assessed in your answer to this question.

Write your answer on separate paper or in your exercise book.

Total [6]

Going for the highest grades

Answer this question if you expect to take the Higher tier test.

4 Explain how a negative feedback system keeps water levels constant in the cells of a human body. In your answer, include the names of the hormone and organs involved in the system.

✎ The quality of written communication will be assessed in your answer to this question.

Write your answer on separate paper or in your exercise book.

Total [6]

Exam tip

When answering 6-mark questions:

- Make sure you know what the question is asking you to do.
- Write down key words to help answer the question.
- Organise the key words in a sensible order.
- Write your answer.
- Check your spelling, punctuation, and grammar.

B3 Life on Earth

1 Make notes about energy transfer between organisms in the box below. Two of the rows have been filled in for you.

- Write the title in the top row.
- Write the most important point (the key idea) in the next row down.
- Write other relevant information in the lower rows.

Title	
Most important point (key idea)	Energy is transferred between organisms when a consumer eats another organism.
Other information	When a consumer eats a plant or animal, only some of the energy is transferred from the plant or animal to the consumer.

2 Do this activity with a friend.

Choose a box from the grid below.

Define the word at the top of the box. Do not use the 'taboo' word.

Get your friend to guess the word you are defining.

Non-living indicators *Taboo words:* • nitrate levels • temperature • carbon dioxide levels	**Mutations** *Taboo words:* • genes • changes • inherited	**Genetic variation** *Taboo words:* • characteristics • inherited • DNA
Competition *Taboo words:* • nutrients • water • survival	**Kingdom** *Taboo words:* • classification • group • animal	**Decomposers** *Taboo words:* • bacteria • dead animals • dead plants
Biodiversity *Taboo words:* • variety • species	**Sustainability** *Taboo words:* • needs • future • environment	**Species** *Taboo words:* • breed • offspring • fertile

3 This question is about the carbon cycle.

a Look at the words in the box.

atmosphere	**photosynthesis**	**combustion**
fossil fuels	**respiration**	**eating**
dying	**decomposing**	

- Write an **s** next to each word that is a carbon store.
- Write a p next to each word that describes a process.

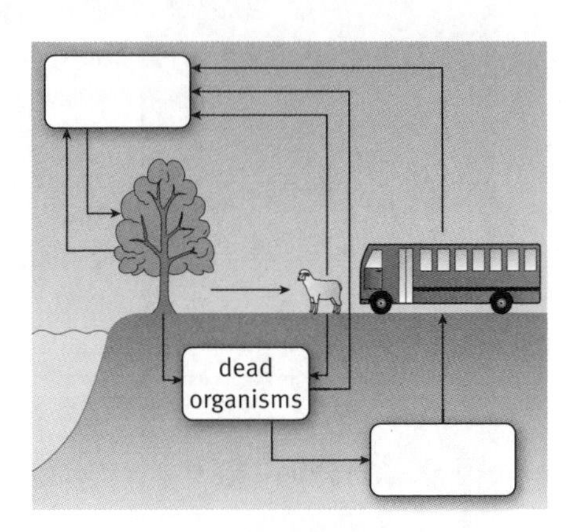

b Write each word in the box above in the correct place on the carbon cycle opposite.

Write the carbon stores in the boxes.

Write process words next to arrows. One arrow does not need a label. You will need to use some process words more than once.

4

The Latin name of one woodlouse species is *Porcellio scaber*.
Woodlice are in these groups:
- invertebrates
- crustaceans
- Animal Kingdom.

Use the information in the box to fill in the gaps below.

The group with the largest number of organisms is the ______________. The group containing organisms with the smallest number of characteristics in common is the ______________. Organisms in the same ______________ can breed to produce ______________ offspring. Within a species, there are fewer organisms with many ______________ in common.

Exam tip

You need to remember what *kingdoms* and *species* are. You will not be expected to remember the names of other types of groups.

B3.1.1–3 What are adaptations?

A **species** is a group of organisms that can breed together to produce **fertile** offspring.

Living organisms have features that help them survive in their environments. These are **adaptations**. Adaptations increase a species' chance of survival by making it more likely that individuals will survive and reproduce.

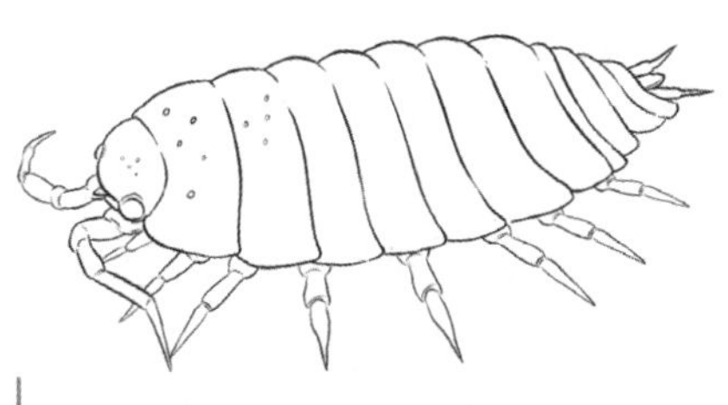

Woodlice live in dark damp places. Their adaptations help them to survive. Their colour makes it hard for predators to see them. Their external skeleton protects the soft parts of their body.

B3.1.4–8 How do species interact?

Within a place where an organism lives – its **habitat** – there is **competition** for resources. Animal species may compete for food or shelter. Plants compete for space and light.

Species in a habitat rely on each other, and on their environment, for food and other needs. The species are **interdependent**. A **food web** shows what eats what in a habitat.

Changes to the food web affect other species. The fox population may decrease. The populations of mice, slugs, beetles, and frogs then may increase. Badger numbers may increase because there are fewer foxes to compete with.

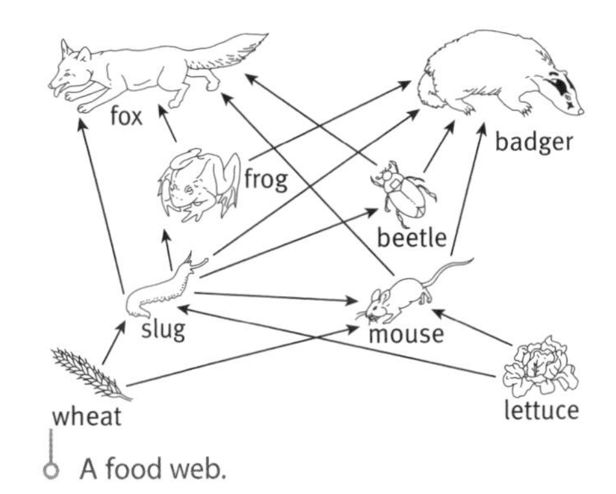

A food web.

B3.1.9–11 How do plants use energy from the Sun?

Plants absorb 1–3% of the light energy from the Sun that falls on their leaves. They use this energy for **photosynthesis**. This produces the chemicals that make up plant cells and store energy. Plants are **producers**.

Animals, bacteria, and fungi depend on plants for food. So nearly all life is dependent on energy from the Sun

B3.1.12–14 How is energy transferred?

Energy is transferred between organisms:

- when animals (**consumers**) eat other organisms
- when decay organisms (**decomposers** and **detritivores**) eat dead organisms and waste materials.

Only a small percentage of the energy at each stage of a food chain is passed on. The rest of the energy:

- is used for life processes (e.g. moving and keeping warm)
- escapes to the surroundings as heat
- is excreted as waste and passed on to decomposers
- cannot be eaten and is passed on to decomposers.

Because so much energy passes out of a food chain, food chains usually have no more than four species.

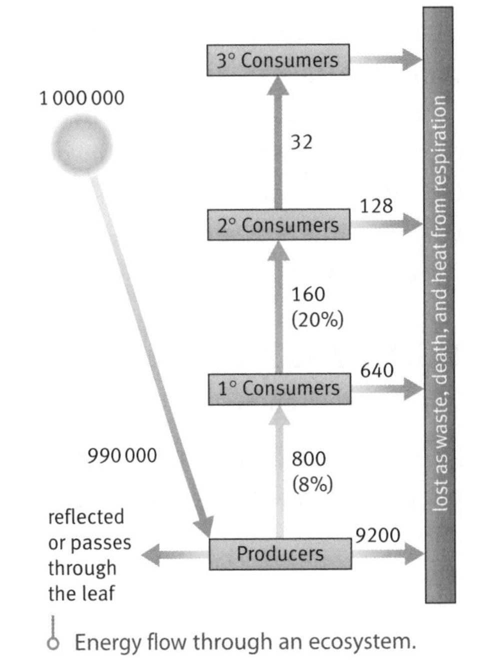

Energy flow through an ecosystem.

You can calculate the percentage efficiency of energy transfer at different stages of a food chain like this:

In the food chain above, 32 units out of 160 units of energy are transferred from secondary consumers to tertiary consumers. So the percentage of the energy transferred is:

$$\frac{32}{160} \times 100 = 20\%$$

B3.1.15–16 How is carbon recycled?

Carbon is recycled through the environment.

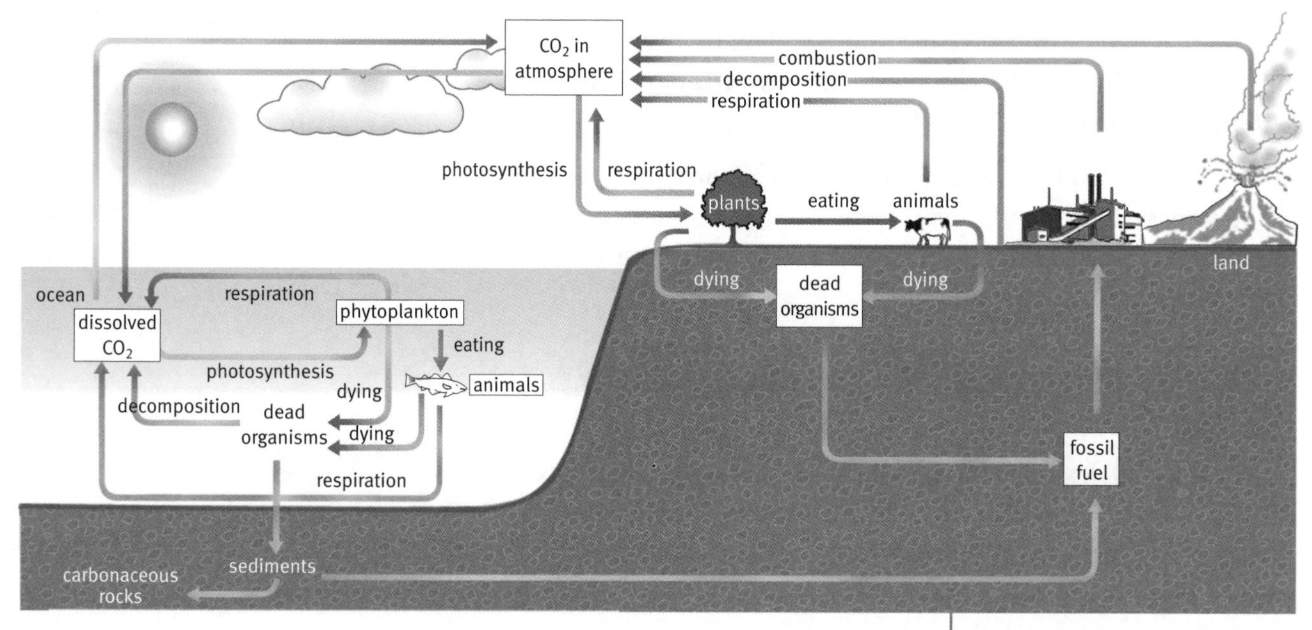

The carbon cycle.

Plants take carbon dioxide out of the atmosphere by **photosynthesis**. This makes glucose. Animals and plants break down glucose in **respiration**. This returns carbon dioxide to the atmosphere. Microorganisms break down the molecules of dead organisms by **decomposition**. The **combustion** of wood and fossil fuels adds carbon dioxide to the atmosphere.

Exam tip

Check the direction of arrows when looking at a carbon cycle. The arrows point to where the carbon atoms are going.

B3.1.17–19 How is nitrogen recycled?

The diagram shows how nitrogen is recycled in the environment.

Microorganisms are vital in the nitrogen cycle. **Decomposer bacteria** break down proteins in dead organisms.

The following processes are part of the nitrogen cycle:

- **Nitrogen-fixing bacteria** in some plant roots convert nitrogen into nitrogen compounds, including nitrates.
- Plants use nitrates to make proteins. Animals digest plant proteins and use them to make animal proteins.
- **Denitrifying bacteria** break down nitrates in the soil and release nitrogen to the air. This is **denitrification**.

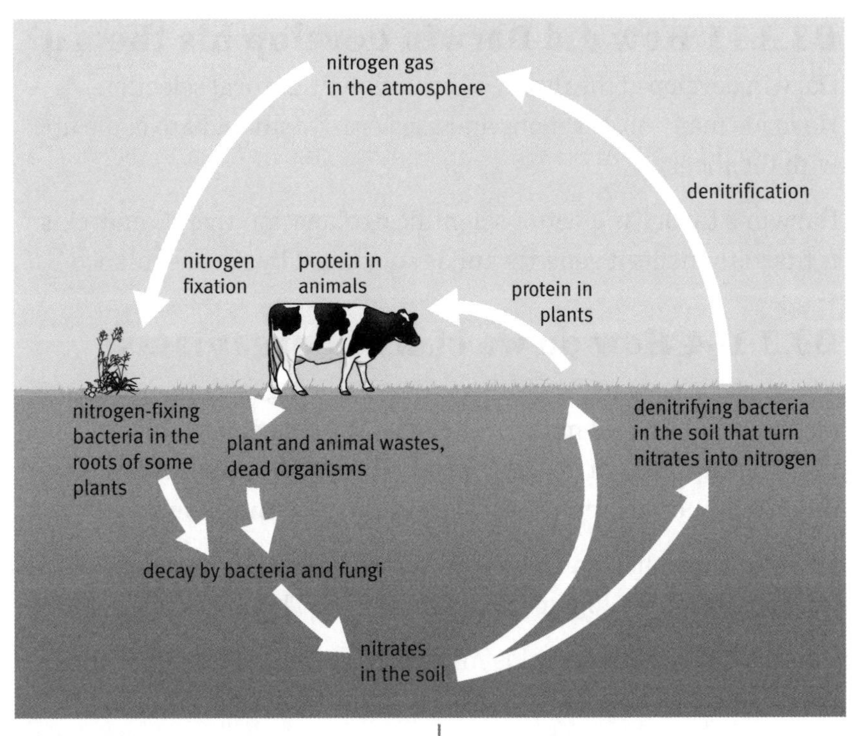

The nitrogen cycle

B3.1.20–22 Monitoring the environment

Scientists use indicators to measure environmental change. **Living indicators** include:

- phytoplankton, to measure ocean temperature changes
- lichens, to monitor air quality
- mayfly nymphs, to monitor oxygen levels in rivers.

Non-living indicators include:

- nitrate levels in streams, rivers, and lakes
- carbon dioxide levels and temperature in air and oceans.

B3.2.1–10 How has life evolved?

Life began about 3500 million years ago. All species evolved from simple living things. Fossils and DNA analysis of living organisms provide evidence for evolution.

There is **variation** between individuals of a species. **Genetic variation** is caused by **mutations** (changes to genes). Mutated genes in sex cells can be passed on to offspring. This occasionally produces new characteristics.

Over time, **evolution** makes species change. New species may develop, too. The changes below may cause evolution:

- mutations and **natural selection**
- environmental changes (some individuals have features that are better suited to a new environment)
- isolation (if a population of a species lives separately from another population of the same species, and changes).

Natural selection is different from **selective breeding**, in which humans choose characteristics for a plant or animal.

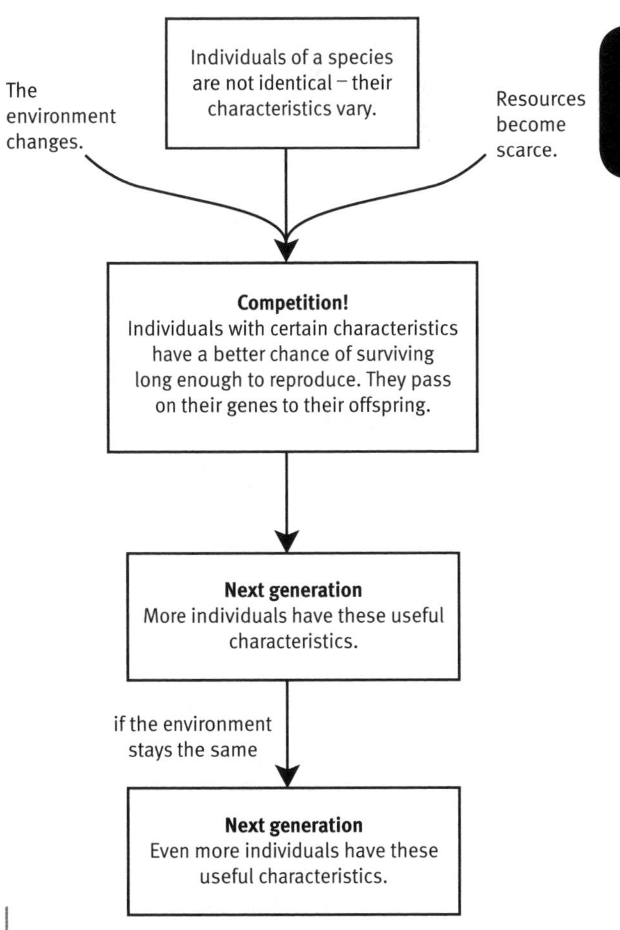

The flow chart shows how natural selection works.

B3

B3.2.11 How did Darwin develop his theory?

Darwin developed the theory of evolution by natural selection. He made many observations and used creative thought to come up with the theory.

Darwin's theory is a better scientific explanation than Lamarck's. It fits with modern genetics and is supported by more evidence.

B3.3.1–4 How do we classify organisms?

There is a huge variety of life on Earth. Living organisms include plants, animals, and microorganisms. Within each of these groups there are millions of species. And within each species there is a much genetic variation. This variety is called **biodiversity**.

Scientists use the similarities and differences of organisms' physical features and DNA to put them into groups. This is **classification**.

Classifying organisms helps make sense of the diversity of life. It also helps to show how organisms have evolved.

Living things have Latin names. The cat is *Felis catus*. It is in these groups:

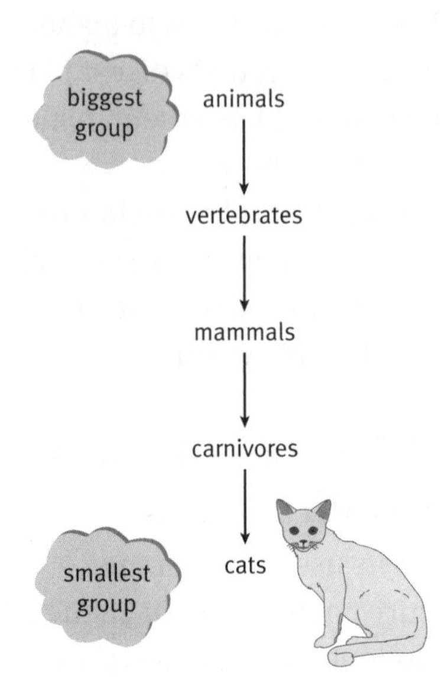

The biggest group that cats belong to is the **Animal Kingdom**. Organisms in the animal kingdom have only a few characteristics in common. The smallest group that cats are in is their species. Its members have many characteristics in common.

B3.3.5–9 Why does biodiversity matter?

Sustainability means meeting the needs of people today without damaging the Earth for people in future. Preserving biodiversity is a vital part of living sustainably:

- We use wild varieties of plant species to develop new varieties of food crops.
- We use plant substances as medicines.

Growing single crops in large fields (**monoculture**) is not sustainable. These crops can easily be attacked by pests and diseases. But crops with more varied alleles are likely to include some resistant plants.

B3.1.8 Why do species become extinct?

If all the members of a species die out, the species is **extinct**. Every year, more species become extinct. A species may become extinct if:

- there are changes in the environment to which a species cannot adapt
- a new species arrives that competes with, eats, or causes disease of the species
- another species in its food web becomes extinct.

B3.3.10–11 Why is packaging a problem?

The production and transport of packaging uses huge amounts of energy. Packaging also creates lots of waste. In landfill sites this waste takes up a lot of space. Biodegradable packaging often fails to decompose in landfill sites because there is not enough oxygen. It is more sustainable to reduce our use of packaging.

Use extra paper to answer these questions if you need to.

1 Write the letter N next to non-living indicators of environmental change. Write the letter L next to living indicators.

- **a** nitrate levels
- **b** lichens
- **c** temperature
- **d** mayfly nymphs

2 Draw lines to match each word to its definition.

Words	Definitions
variation	the place where an organism lives
mutations	everything around an organism, including air, water, and other living things
habitat	differences between organisms
environment	changes to genes

3 Choose words from the box to fill in the gaps. The words may be used once, more than once, or not at all.

small chemicals dead heat warm large materials respiration photosynthesis energy moving waste

Plants absorb a __________ percentage of the Sun's energy for __________. This energy is stored in the __________ of plant cells. When one organism eats another, only about 10% of the __________ is transferred to the organism. This is because some of the energy is transferred in life processes such as __________ and keeping __________, and some is transferred to the surroundings as __________. Also, some energy remains in undigested __________. In the same way, when decomposers feed on __________ organisms and waste __________, only some of the energy is transferred to the decomposers.

4 Highlight the statements below that are **true**.Then write corrected versions of the statements that are **false**.

- **a** Life on Earth began about 3500 billion years ago.
- **b** Mutated genes in sex cells cannot be passed on to offspring.
- **c** In selective breeding, humans choose individual plants or animals to breed from.
- **d** Evolution is the process by which species gradually change over time.

5 Polar bears have thick white fur. They are good swimmers. Explain how these adaptations increase their chance of survival in the arctic.

6 The diagram below shows part of a food chain in Australia.

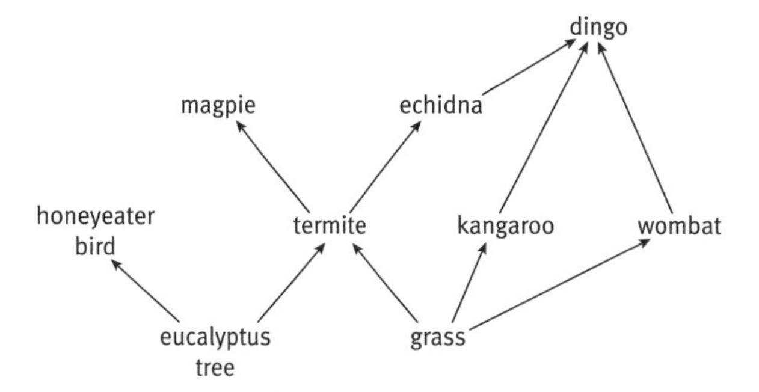

- **a** Name the animals in the food web that eat grass.
- **b** Name the producers in the food web.
- **c** Name the primary consumers in the food web.
- **d** Explain what might happen to the wombat population if the kangaroo population decreases.
- **e** Explain what might happen to the honeyeater bird population if eucalyptus trees are cut down.
- **f** Explain what might happen to the wombat population if dingoes die.

7 In the food chain below, there are 200 units of energy in all the herrings eaten by a shark. The amount of energy transferred to the shark is 28 units. Calculate the percentage of energy that is transferred.

phytoplankton ⟶ zooplankton ⟶ herring ⟶ shark

8 List three ways by which changes in the environment may cause a species to become extinct.

9 Explain why Darwin's theory of evolution by natural selection is a better theory than Lamarck's explanation of evolution.

H

10 Explain the meaning of the word *interdependence*.

11 What is a detritivore?

12 Draw lines to link the name of each process in the nitrogen cycle to what it does.

Name of process	What it does
decay	breaks down nitrates to form nitrogen
nitrogen fixation	forms nitrates from nitrogen
excretion	breaks down proteins
denitrification	removes waste from an organism

B3

1 Read the information in the box.

Here is part of an Antarctic food web:

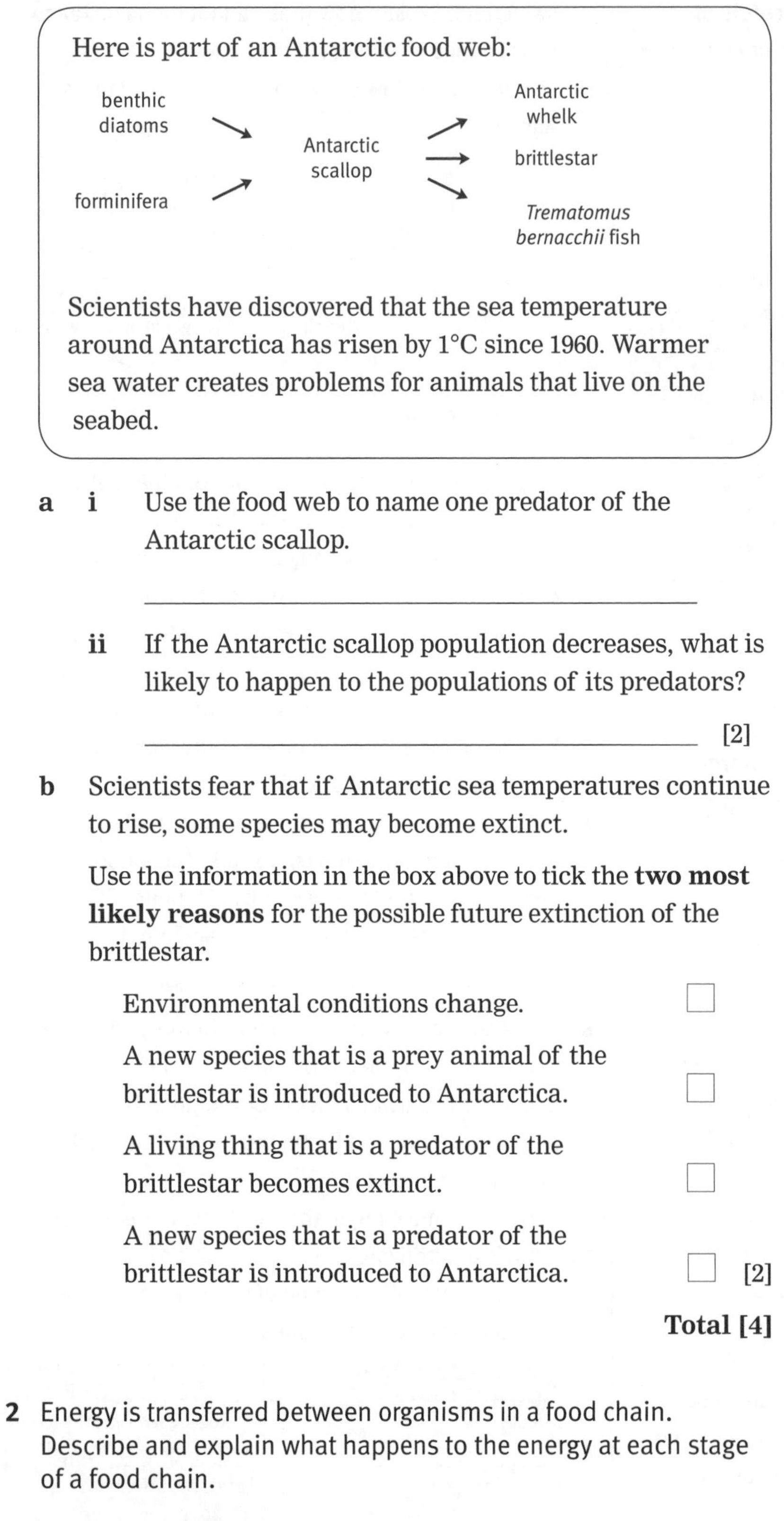

Scientists have discovered that the sea temperature around Antarctica has risen by 1°C since 1960. Warmer sea water creates problems for animals that live on the seabed.

a **i** Use the food web to name one predator of the Antarctic scallop.

____

ii If the Antarctic scallop population decreases, what is likely to happen to the populations of its predators?

____ [2]

b Scientists fear that if Antarctic sea temperatures continue to rise, some species may become extinct.

Use the information in the box above to tick the **two most likely reasons** for the possible future extinction of the brittlestar.

- Environmental conditions change. ☐
- A new species that is a prey animal of the brittlestar is introduced to Antarctica. ☐
- A living thing that is a predator of the brittlestar becomes extinct. ☐
- A new species that is a predator of the brittlestar is introduced to Antarctica. ☐ [2]

Total [4]

Exam tip

Practise predicting how changes in the population of one species in a food chain will affect the populations of other species in the food chain.

2 Energy is transferred between organisms in a food chain. Describe and explain what happens to the energy at each stage of a food chain.

✎ The quality of written communication will be assessed in your answer to this question.

Write your answer on separate paper or in your exercise book.

Total [6]

3 Scientists have studied how the cat family evolved. They discovered that lions and domestic cats shared a common ancestor 10.8 million years ago.

a How might the scientists have obtained evidence to support their explanation?

Tick the two best answers.

studying fossils ☐

analysing the fur of cat ancestors ☐

analysing the DNA of modern cats, lions, and other species of the cat family ☐

analysing the blood of cat ancestors ☐ [2]

b i Complete this sentence.

Domestic cats and lions evolved partly as a result of a process called natural ____________. [1]

ii The stages below explain how evolution made changes to one species: lions.

The stages are in the wrong order.

A More individuals in this generation had features that helped them survive in their new environment.

B Early lions migrated from Asia to Africa. Some individuals had features that helped them survive in the new environment.

C Individual lions are not identical; the species shows variation.

D These lions bred. They passed on their genes to their cubs.

Fill in the boxes to show the correct order. The first one has been done for you.

C			

[2]

Going for the highest grades

H

c The diagram shows scientists' ideas about when some species of the cat family began evolving from their common ancestor.

B 3

For example, the domestic cat and the ocelot last shared a common ancestor 8.0 million years ago.

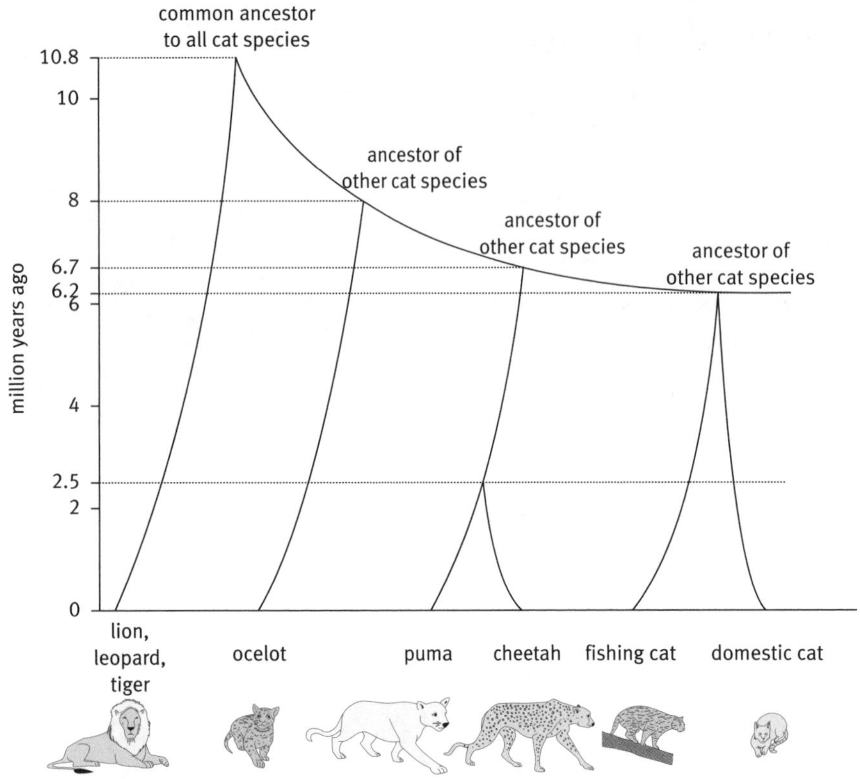

i When did the cheetah and the domestic cat last share a common ancestor?

______________________________ [1]

ii Which species on the chart probably has DNA that is most similar to that of the domestic cat?

______________________________ [1]

iii Read the information in the box opposite.

Complete the following sentences. Choose words from this list.

natural	**artificial**	**survival**
selection	**environmental**	**isolation**

This shows how ______________ changes and ______________ help to produce new species.

Changes to genes (mutations) and ______________ ______________ also help to produce new species. [3]

The common ancestor of the puma and cheetah lived in North America. Individuals of the ancestor species migrated. Some went to South America and evolved into a new species: the puma. Others went to Africa and evolved to become a different species: the cheetah.

Total [10]

B4 The processes of life workout

1 Use words from the box below to label the cell diagrams.

nucleus	**cytoplasm**	**cell membrane**	**mitochondrion**
cell wall	**chloroplast**	**circular DNA**	**vacuole**

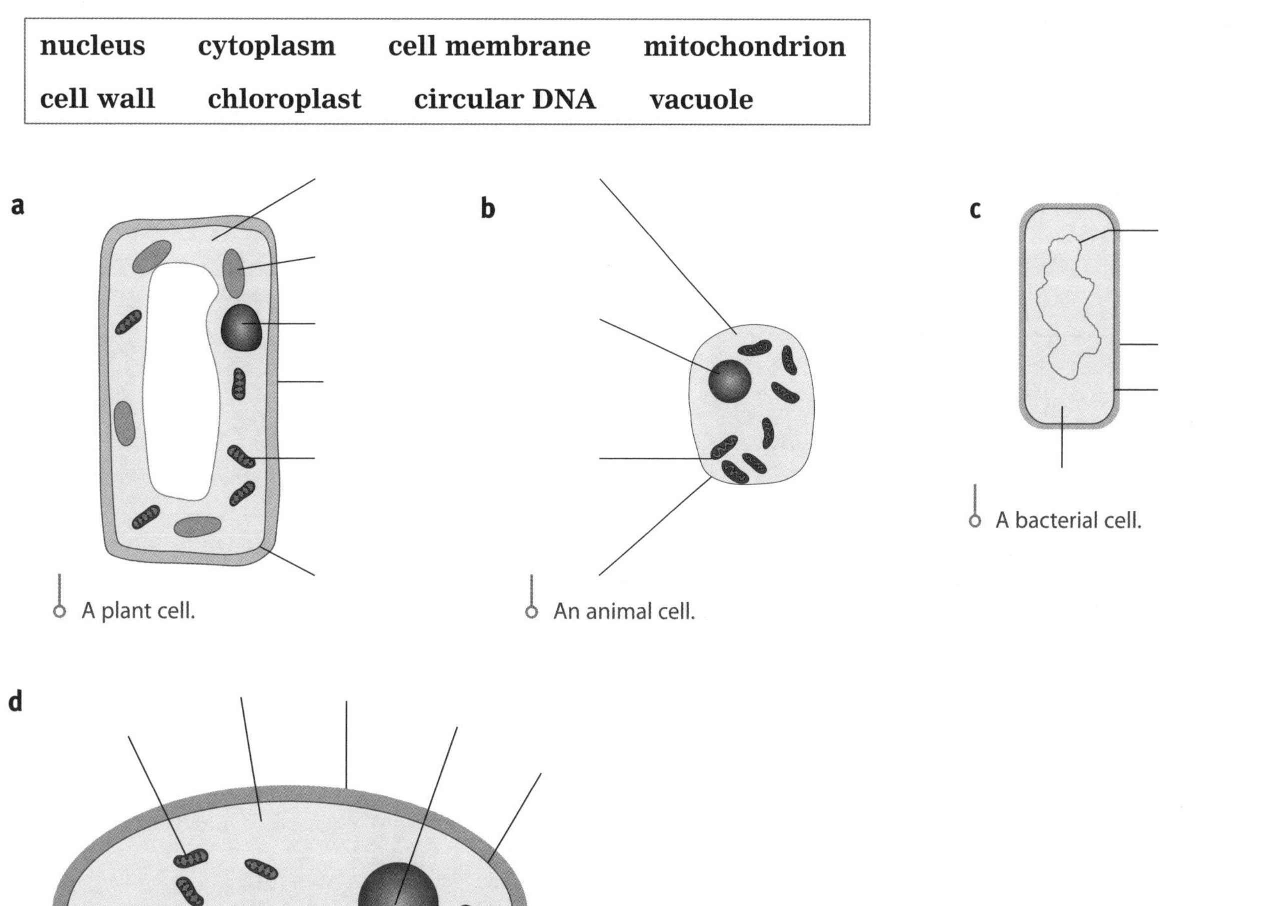

A plant cell.

An animal cell.

A bacterial cell.

A yeast cell.

2 The stages opposite show how an enzyme catalyses a reaction at its active site.

The reaction involves breaking down one molecule to make two smaller molecules.

Write the letters of the stages in the correct order in the boxes below.

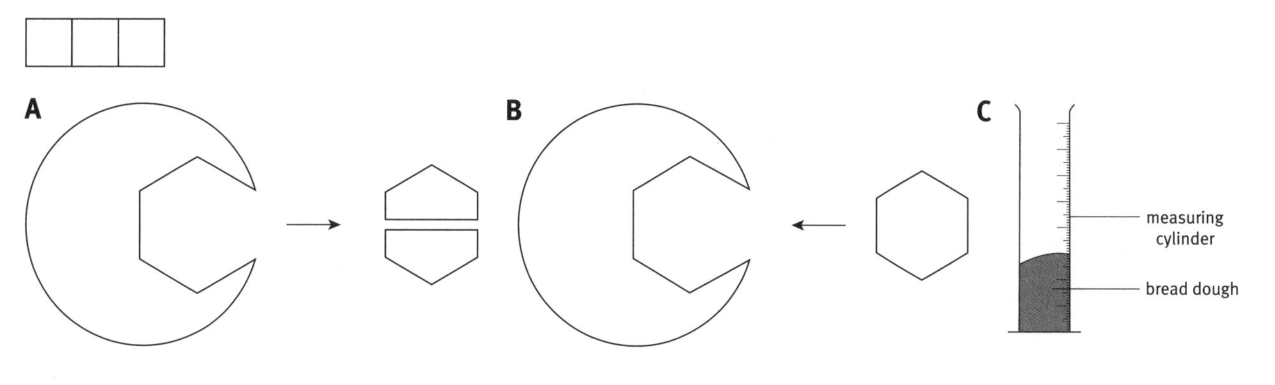

3 The graph shows the link between pH and rate of reaction for two different reactions that are controlled by enzymes.

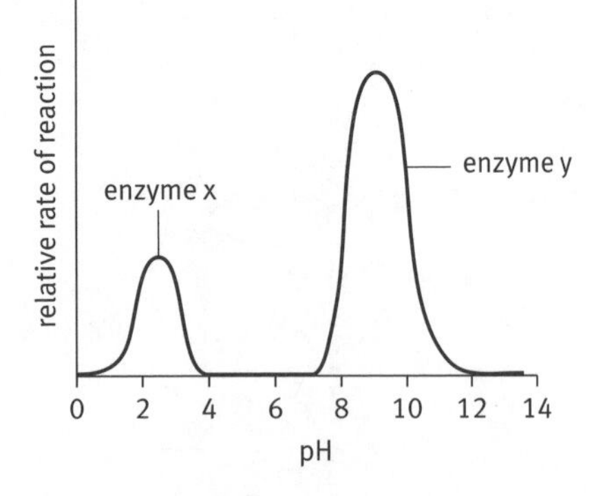

a Which enzyme works best in acidic conditions? ________

b What is the pH range in which enzyme X is active? ______

c Would enzyme Y work in a human stomach? Explain your answer.

____

4 Gareth set up the investigation below.

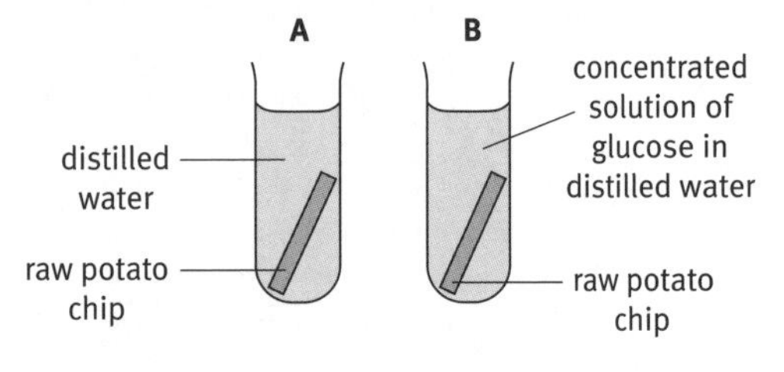

Write the letter of one or more test tubes next to each statement below.

a This chip gets bigger. _____

b This chip feels firmer when it is taken out of the test tube. _____

c Overall, more water has moved from the surrounding liquid into the chip. _____

d Overall, more water has moved out of the chip into the surrounding liquid. _____

e Water gets into and out of the potato by diffusion. _____

f Water gets into and out of the potato by osmosis through cell membranes. _____

B4.1.1, B4.1.4 Why are there chemical reactions in cells?

All living things carry out seven **life processes**. Living things move, reproduce, sense their surroundings, grow, excrete waste products, and need nutrition.

These processes depend on chemical reactions which happen in cells. Many of the reactions need energy. This energy is released by the seventh life process – **respiration**. Respiration is a series of chemical reactions that releases energy by breaking down large food molecules in all living cells.

B4.1.5–7 What are enzymes?

Enzymes speed up chemical reactions in cells. An enzyme is a natural **catalyst**. Catalysts speed up chemical reactions without getting used up.

Enzymes are proteins. They are made in cells. In cells, genes carry the instructions to make enzymes.

Every reaction in a cell needs its own enzyme. The reacting molecules must fit exactly into the **active site** of the enzyme. So every enzyme has its own shape. This is the **lock and key model**.

B4.1.8–11 What conditions do enzymes need?

Enzymes work best at their **optimum temperature**.

- Below the optimum temperature, reactions are slow.
- If the temperature is too high, an enzyme changes its shape. It no longer catalyses its reactions.

The enzyme has been **denatured**.

FAST
SLOW
rate of reaction
temperature (°C)
0 10 20 30 40 50 60
As the temperature rises, there are more successful reactions.
At the enzyme's optimum temperature, there is a high rate of reaction.
At low temperatures, the rate of reaction is low.
At high temperatures the enzyme's active site is changed. Molecules cannot fit into the active site. The rate of reaction slows.

The graph shows how the rate of an enzyme-catalysed reaction changes with temperature.

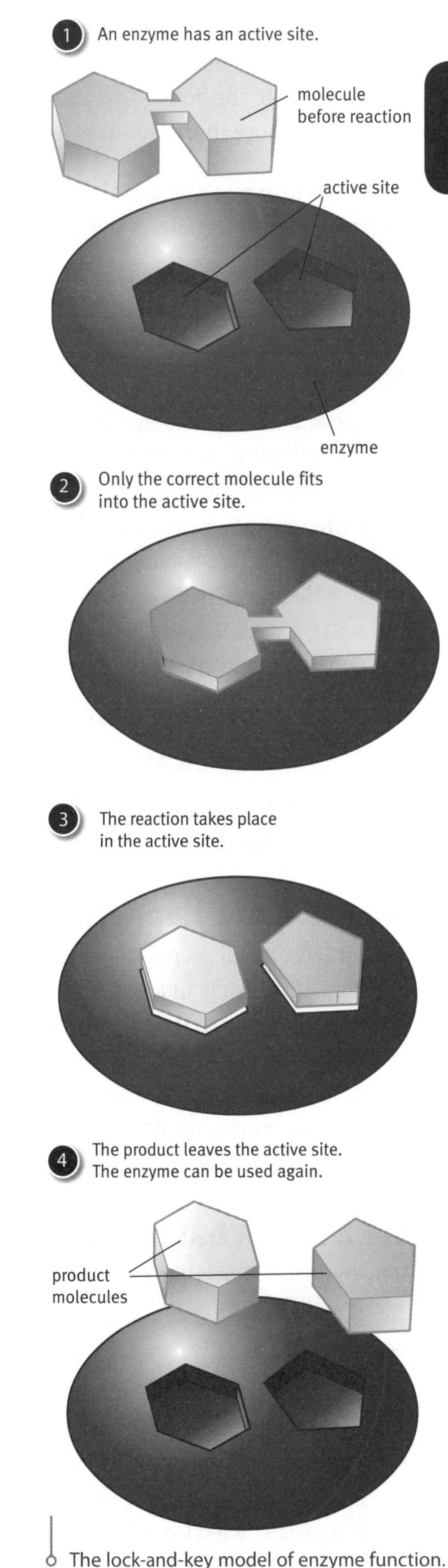

The lock-and-key model of enzyme function.

H An enzyme's activity at different temperatures is a balance between:
- increasing reaction rates as temperature increases
- changes to the active site, including denaturing, as the temperature increases above the optimum.

Enzymes also work best at their optimum pH. For example, in the stomach, pepsin breaks down proteins most efficiently at pH 2.

H At other pH values, the shape of pepsin's active site changes. The protein molecules it breaks down no longer fit into its active site.

B4.1.2–3, B4.2.1–2 What is photosynthesis?

At the start of every food chain is a plant. Plants use energy from sunlight to make their own food by **photosynthesis**. Photosynthesis happens like this:
- **Chlorophyll**, a green pigment, absorbs light energy.
- In plant cells, this energy brings about a series of reactions in which carbon dioxide and water molecules join together to make glucose, a sugar.
- Oxygen is the waste product of the reactions.

$$\text{carbon dioxide} + \text{water} \xrightarrow{\text{light energy}} \text{glucose} + \text{oxygen}$$

H
$$6CO_2 + 6H_2O \xrightarrow{\text{light energy}} C_6H_{12}O_6 + 6O_2$$

Photosynthesis also happens in some microorganisms, for example, in phytoplankton.

B4.2.4 What happens to the glucose?

The glucose made in photosynthesis has three main uses:
- Some glucose is made into the chemicals that plant cells need to grow, for example, chlorophyll, proteins, and **cellulose**.
- Some glucose molecules join together to make starch molecules. Starch is a storage chemical. It breaks down into glucose molecules when the plant needs more glucose.
- Some glucose is used in respiration.

B4.2.5–6 What's in a plant cell?

The diagram shows the structure of a typical plant cell.

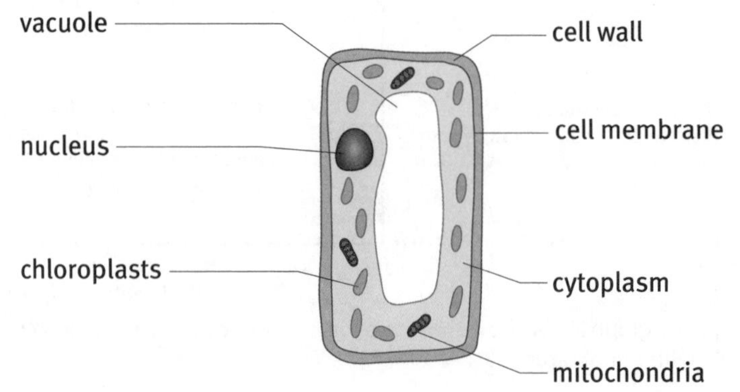

Each part of a plant cell has a vital function:
- **Chloroplasts** contain chlorophyll, and the enzymes for photosynthesis.
- The **nucleus** contains DNA. This carries the genetic code for making enzymes and other proteins needed for photosynthesis.

- Enzymes and other proteins are made in the **cytoplasm**.
- The **cell membrane** lets gas and water molecules into the cell, but stops chemicals with bigger molecules getting out.
- Respiration happens in **mitochondria**.
- The **vacuole** contains glucose molecules, dissolved in water.
- The **cell wall** is rigid. It is made of cellulose.

B4.2.7–13 How do chemicals get into cells?

Molecules get into and out of cells by **diffusion**. In diffusion, molecules move from a region of higher concentration to a region of their lower concentration. Diffusion is a passive process – it does not need extra energy.

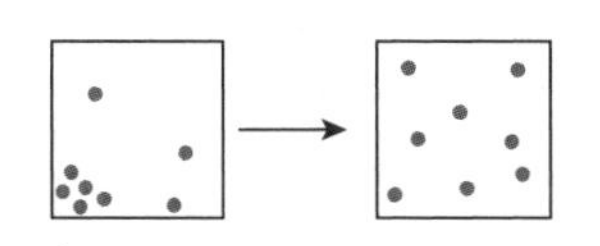

In diffusion, particles spread out from where there are lots of them to where there are fewer of them.

Carbon dioxide molecules diffuse into leaves through tiny pores. Oxygen molecules diffuse out of leaves through these same pores.

Water gets into and out of a cell through its **partially permeable membrane**. Small molecules can get through the membrane, but bigger ones cannot. Water molecules diffuse through the membrane from a dilute solution (where there are more water molecules) to a more concentrated solution (where there are fewer.) This type of diffusion is **osmosis**. Water moves into root cells by osmosis.

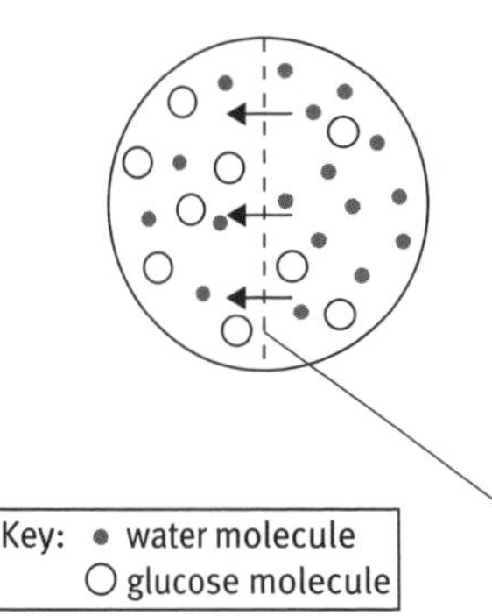

In osmosis, water molecules move through a partially permeable membrane.

Plants take in minerals such as nitrates through root cells. They use nitrogen atoms from nitrates to make proteins.

Active transport moves nitrate ions across cell membranes. Energy from respiration helps transport the particles from a region of lower concentration outside the cell to a region of higher concentration inside the cell. The diagrams summarise active transport.

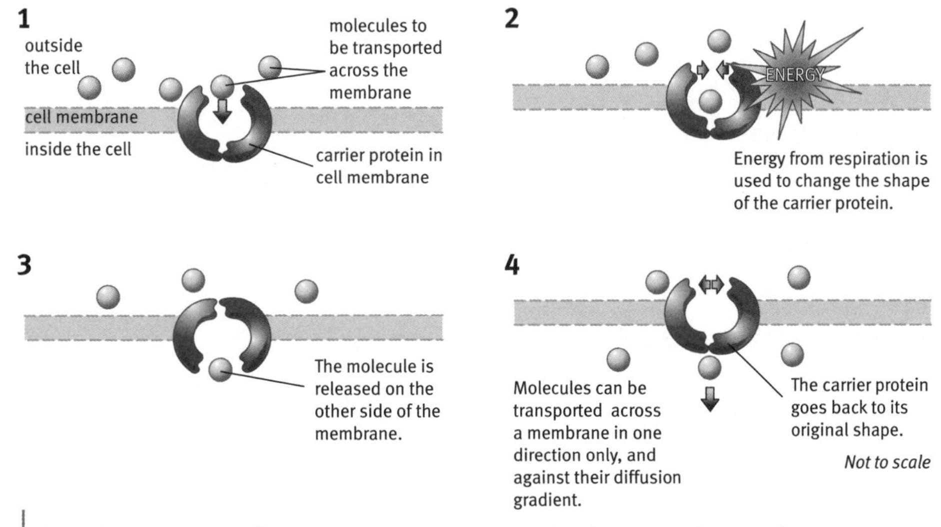

In active transport, cells use energy to transport molecules across the membrane.

B4.2.14–15 What speeds up photosynthesis?

The rate of photosynthesis depends on several factors, including:

- **Temperature** – photosynthesis is faster at higher temperatures.
- **Carbon dioxide concentration** – increasing this speeds up photosynthesis.
- **Light intensity** – at low light levels, increasing the amount of light increases the rate of photosynthesis. But above a certain point, increasing the amount of light no longer increases the rate of photosynthesis. Some other factor, for example, carbon dioxide concentration, now limits the rate. This is the **limiting factor**.

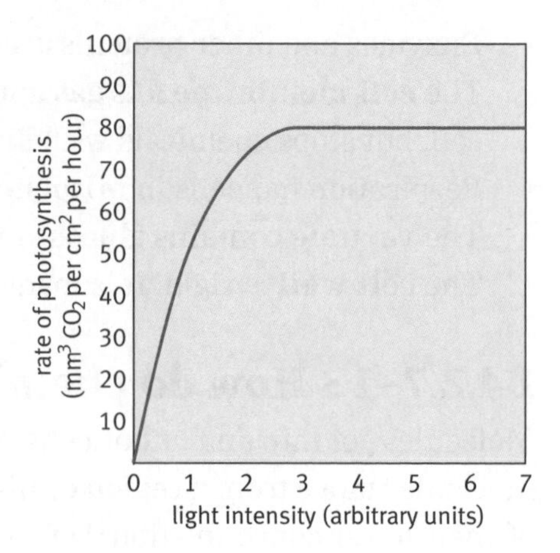

B4.2.16–17 How can we investigate plants and light?

Different plants are adapted to different light levels. You can investigate the plants in different habitats using:

- a **light meter** to measure light intensity
- a **quadrat** (placed at random) and **identification key** to survey the plants in a square metre.

To investigate how plant species change gradually from one area to another, take samples at intervals along a straight line, or **transect**.

B4.3.1–2 Why do living organisms need energy?

Living organisms use energy from respiration for some chemical reactions in cells, including those involved in:

- movement
- active transport (H)
- synthesising (making) big molecules, including:
  - polymers (such as starch and cellulose) from glucose in plants
  - amino acids (from glucose and nitrates) to make proteins in plant, animal, and microbial cells.

B4.3.3–5 What is aerobic respiration?

Aerobic respiration releases energy. It happens in plant and animal cells, and in some microorganisms. It needs oxygen.

glucose + oxygen ⟶ carbon dioxide + water

(H) $$C_6H_{12}O_6 + 6O_2 \longrightarrow 6CO_2 + 6H_2O$$

B4.3.6–9, 4.3.12 What is anaerobic respiration?

Anaerobic respiration releases less energy per glucose molecule than aerobic respiration. It happens in plant and animal cells, and in some microorganisms, when there is little or no oxygen, for example:

- in plant roots, when the soil is waterlogged
- in bacteria, in deep puncture wounds
- in human muscle cells during vigorous exercise.

This equation summarises anaerobic respiration in animal cells, and in some bacteria:

$$\text{glucose} \longrightarrow \text{lactic acid}$$

In plant cells, and in some microorganisms such as yeast, anaerobic respiration forms different products:

$$\text{glucose} \longrightarrow \text{ethanol} + \text{carbon dioxide}$$

Anaerobic respiration in yeast produces carbon dioxide, which makes bread rise. The process also makes ethanol, for alcoholic drinks.

Some bacteria produce methane gas when they break down organic materials such as manure. This **biogas** is a useful fuel.

B4.3.10–11 What's in a cell?

The diagrams show the structures of animal, bacteria, and yeast cells.

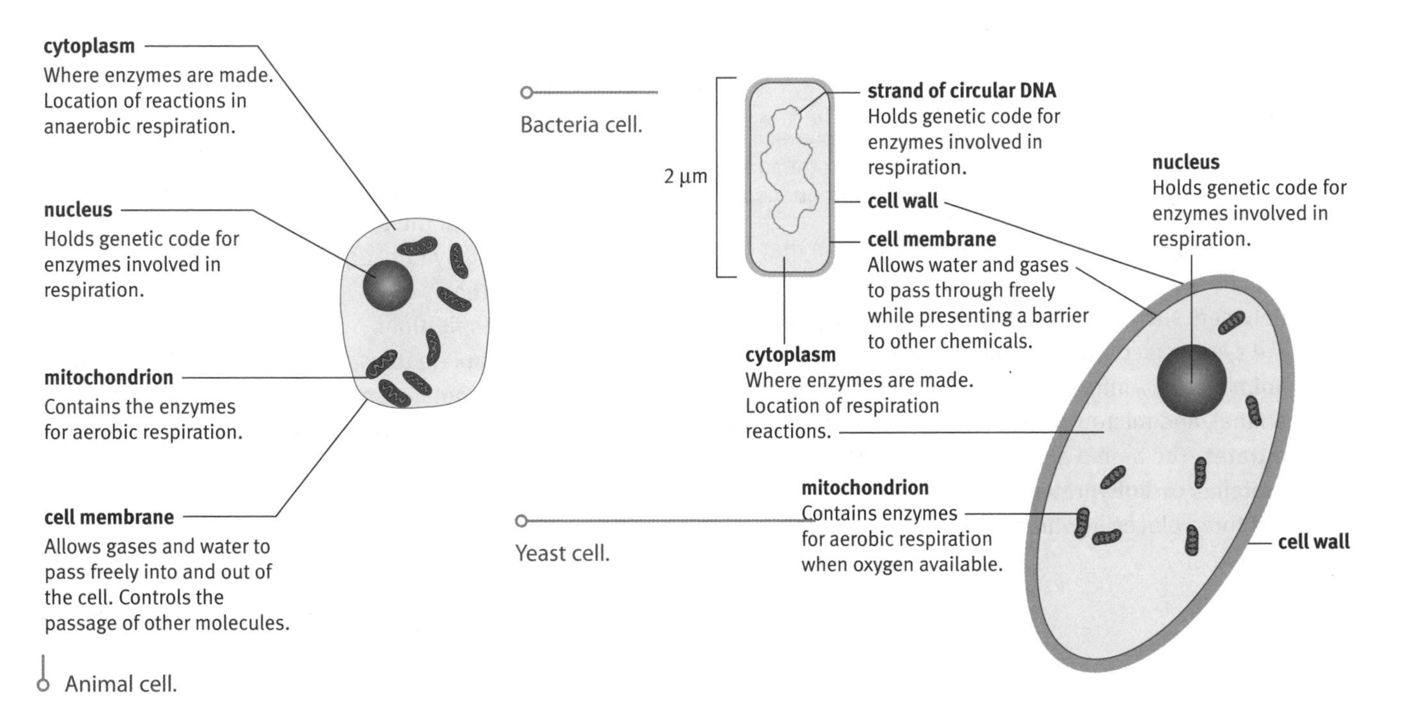

Animal cell.

Bacteria cell.

Yeast cell.

Use extra paper to answer these questions if you need to.

1 List the seven processes of life.

2 Write **T** next to the statements below that are true. Write corrected versions of the statements that are false.

a The waste product of photosynthesis is carbon dioxide.

b In aerobic respiration, glucose and oxygen react together to make carbon dioxide and water.

c Anaerobic respiration releases more energy per glucose molecule than aerobic respiration.

d Enzymes are carbohydrates that speed up chemical reactions.

e Photosynthesis happens in plant and phytoplankton cells.

3 Copy and complete the word equations below.

a Photosynthesis:
carbon dioxide + water ⟶ ___ + ___

b Aerobic respiration:
glucose + ___ ⟶ ___ + ___

c Anaerobic respiration in animal cells:
glucose ⟶ ___

d Anaerobic respiration in yeast and plant cells:
glucose ⟶ ___ + ___

4 Match each word to its definition.

Word	Definition
diffusion	making a chemical with bigger particles from ones with smaller particles
osmosis	movement of molecules from a region of their higher concentration to one of their lower concentration
synthesis	movement of water through a partially permeable membrane to a region of their higher concentration to one of their lower concentration

5 In each pair of bold words, highlight the one that is correct.
Polymers such as **starch/oxygen** and **water/cellulose** are synthesised from **glucose/nitrates** in **plant/animal** cells. Amino acids are synthesised in plant, animal, and microbial cells from glucose and **starch/nitrates**. The amino acids join together to make **proteins/carbohydrates**.

6 List three places in which anaerobic respiration happens.

7 Kish measures the rate of an enzyme-catalysed reaction at different temperatures. His results are on the graph below.

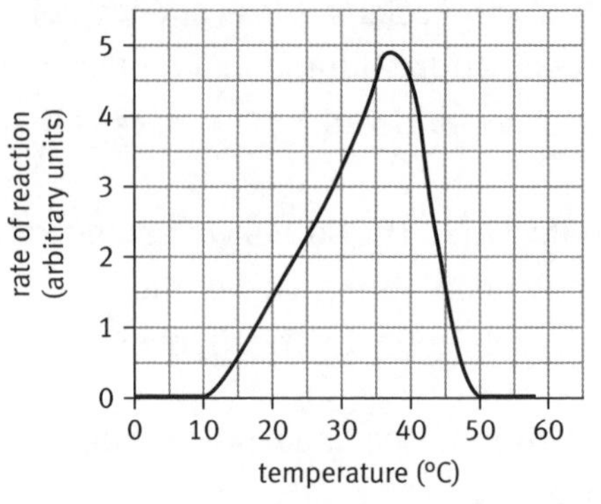

a What is the rate of reaction at 20 °C?

b What is the optimum temperature for the enzyme?

c Over what range of temperatures does the enzyme catalyse the reaction?

d Why does the enzyme not catalyse the reaction above a certain temperature?

8 A scientist recorded the rate of photosynthesis of a plant over 24 hours. She plotted her results on the graph below.

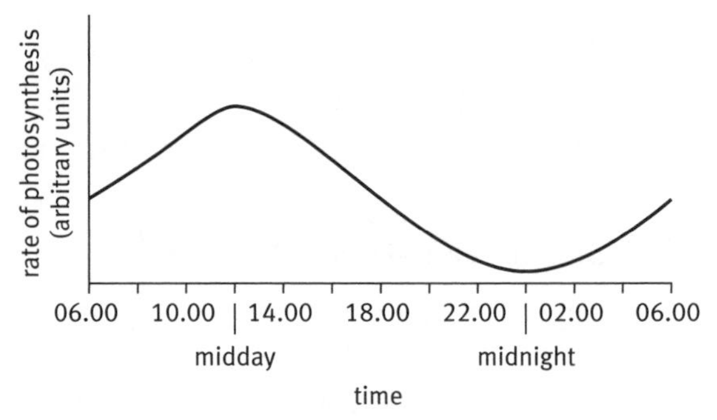

a At what time was the rate of photosynthesis lowest?

b At what time was there the highest concentration of oxygen in the air just around the leaves?

c At what time was most carbon dioxide being removed from the air?

H 9 Write balanced symbol equations to summarise:

a photosynthesis

b aerobic respiration.

10 Copy and complete the sentences below. Use words from this list: roots, respiration, nitrates, energy.
In active transport, chemicals such as _______ are absorbed by plant _______. The process requires _______. This comes from the process of _______.

1 Riana investigates anaerobic respiration in yeast.

She mixes yeast with sugar and warm water.

After 10 minutes, she adds the yeast mixture to flour, and kneads the dough.

She then sets up the apparatus opposite.

Riana measures the height of the dough every 5 minutes for 25 minutes.

The temperature of the room was 24 °C.

Her results are in the table.

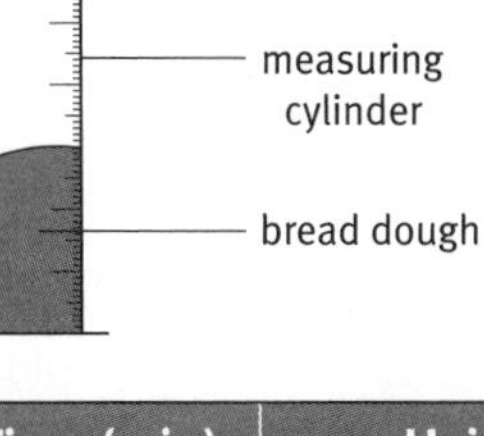

Time (min)	Height of dough (cm)
0	5
5	10
10	13
15	15
20	16
25	16

a Describe the pattern shown by the results.

____

______________________________ [2]

b Riana discusses the investigation with other students.

Here are their ideas.

Zita: The higher the temperature, the faster the dough will rise.

Tom: The dough would rise faster if you added alkali to the mixture of yeast, sugar, and water to increase the pH.

Sam: The dough would rise faster if you put it under a lamp.

i Is Zita's idea correct? Explain your answer.

____

____

______________________________ [3]

ii Is Sam's idea correct? Explain your answer.

____

______________________________ [2]

iii Suggest how Tom could test his idea.
Include any safety precautions he would need to take, and the results he would expect if his idea was correct.

____

____

____

______________________________ [4]

c Write a word equation for the anaerobic respiration reaction that happens in yeast.

______________________________ [1]

Exam tip

When you are asked to explain an answer, make sure you give detailed reasons to back up your response.

d Draw lines to match each cell structure with its function in respiration.

Structure
nucleus
cytoplasm
mitochondria

Function
The enzymes used in aerobic respiration are found here.
The enzymes used in anaerobic respiration are found here.
The genetic code for making enzymes used in respiration is found here.

[2]

Total [14]

2 Describe and explain how substances move in and out of different types of plant cell by diffusion, including osmosis.

✎ The quality of written communication will be assessed in your answer to this question.

Write your answer on separate paper or in your exercise book.

Total [6]

3 Artem investigates the effect of light on the rate of photosynthesis.

He sets up the apparatus opposite.

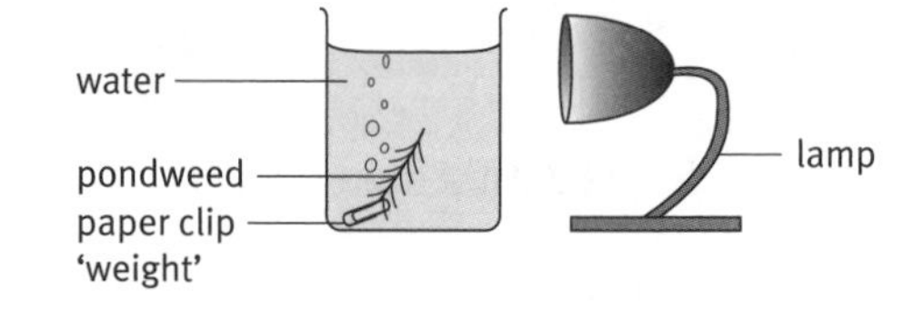

The temperature is 18 °C.

Artem counts the number of bubbles released by the pondweed in one minute.

His results are in the table.

Distance of plant from lamp (cm)	Number of bubbles in one minute			
	Test 1	Test 2	Test 3	Mean
100	10	16	13	13
80	15	18	18	17
60	20	20	23	21
40	35	37	33	35
20	45	49	53	49

a **i** Give the range for the number of bubbles when the distance from the lamp was 80 cm.

____________________________ [1]

ii Plot the mean results on a graph. Use the axes opposite. [2]

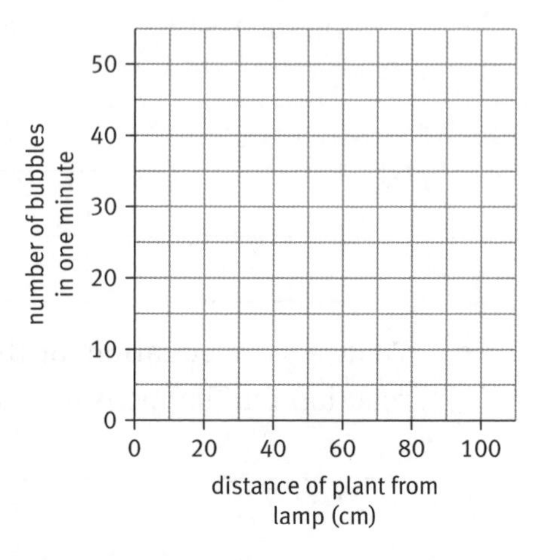

b Write a conclusion for the investigation.

____

______________________________ [2]

c Outline how Artem could use the apparatus above to investigate the effect of temperature on the rate of photosynthesis.

Include the names of any extra pieces of apparatus he would need.

____

____

______________________________ [2]

d **i** Write a word equation to summarise photosynthesis.

______________________________ [1]

ii Name the structure in a plant cell that contains the enzymes for the reactions in photosynthesis.

______________________________ [1]

Total [9]

4 Katya and Martha did some fieldwork to investigate the effect of light on plants growing in a wildlife park.

They used a quadrat and an identification key to survey the plants growing in different square-metre sections of the wildlife park.

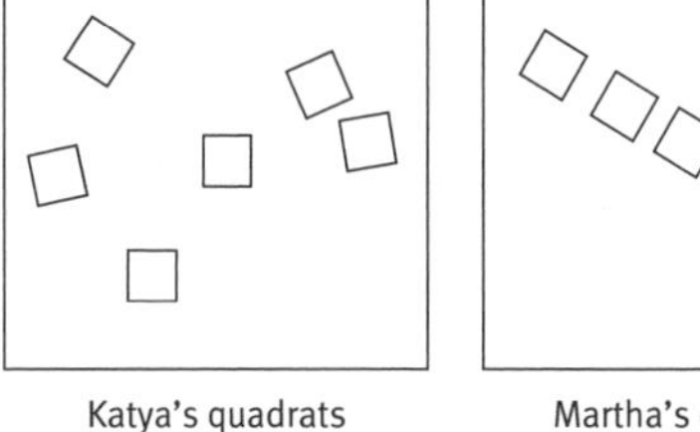

Katya's quadrats

Martha's quadrats

a Katya and Martha drew diagrams to show where they placed their quadrats. Who was studying a transect? Suggest why she decided to study a transect.

____

____

______________________________ [1]

b The diagrams show what Martha found in three of her quadrats.

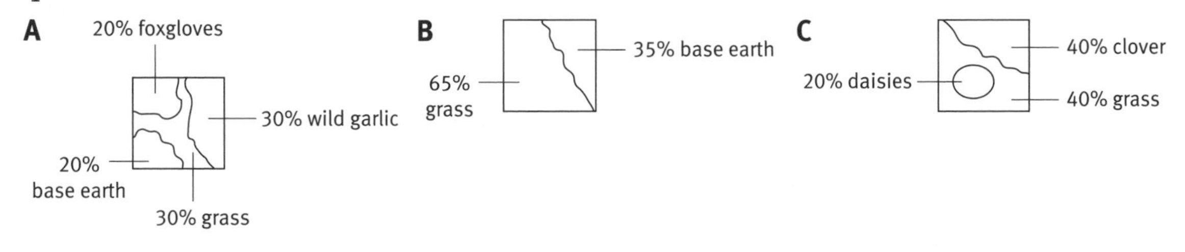

Martha also made some notes from a flower identification key:

Which quadrat above (**A**, **B**, or **C**) is most likely to have been placed under some trees?

Give a reason for your answer.

____

____ [1]

Foxglove – likes shade
Clover – grow well in full sunlight
Daisies – like full sunlight
Wild garlic – likes shade

Total [2]

Going for the highest grades

5 Green plants make their own food by photosynthesis.

a Write a balanced symbol equation to summarise photosynthesis.

____ [2]

b The graph shows the relationship between light intensity and the rate of photosynthesis at two different concentrations of carbon dioxide.

rate of photosynthesis
0.4% CO_2 concentration
0.04% CO_2 concentration
light intensity

Explain what the graph shows about how two factors – light intensity and carbon dioxide – affect the rate of photosynthesis.
Include ideas about limiting factors in your answer.

____

____

____ [3]

Exam tip

If three marks are available, try to make three points in your answer.

c Plants take in nitrates through their roots.

i Describe what the nitrates are used for.

____

____ [2]

ii Describe how plant root hair cells absorb nitrates.

____

____

____ [5]

Total [2]

B5 Growth and development workout

1 The flow diagram is about mitosis and meiosis in humans.

Use words and numbers from the box to fill in the gaps.
Use each word or number once, more than once, or not at all.
The pictures in the flow diagram may help you.

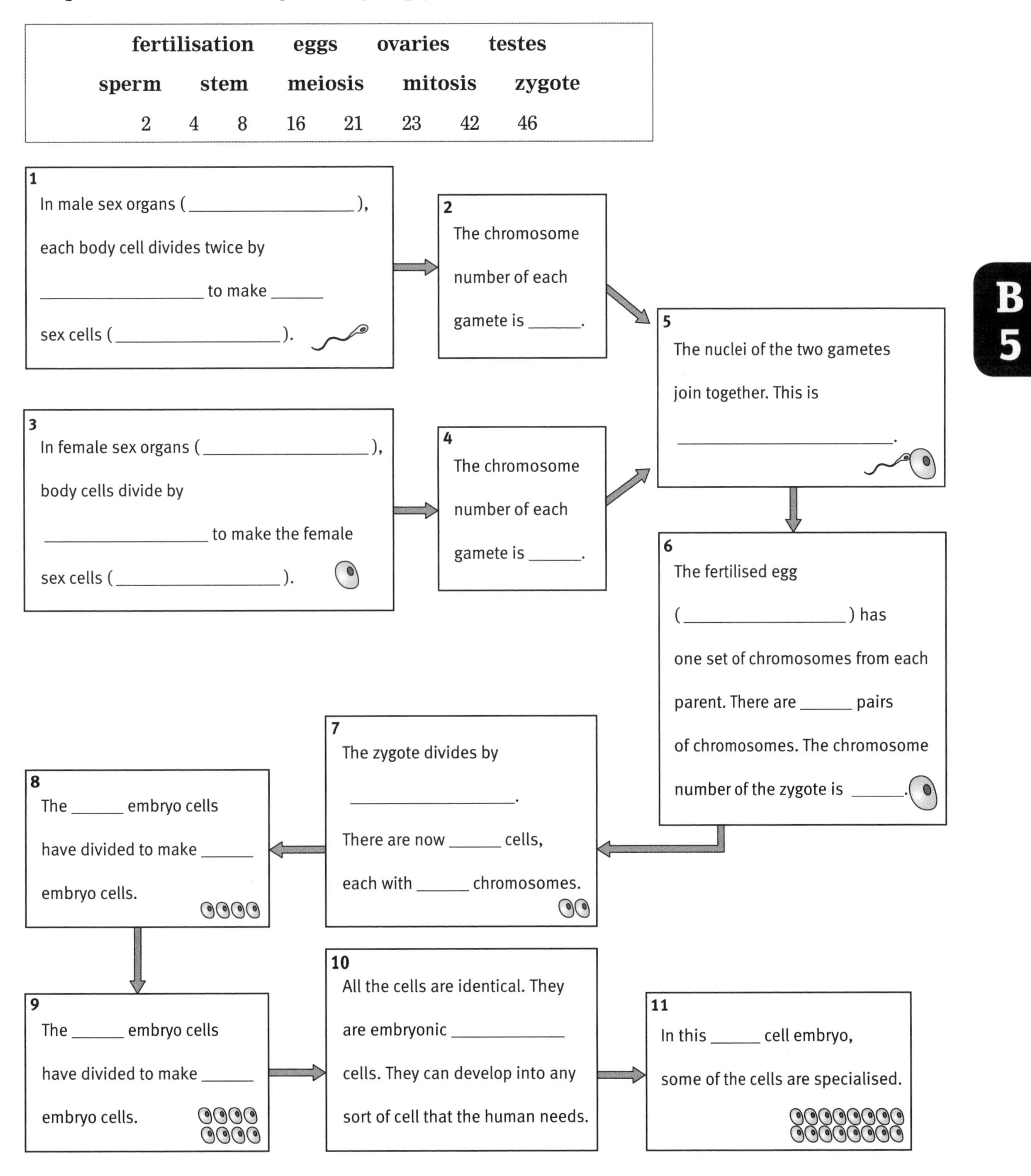

2 All the words in this wordsearch are about the growth and development of plants and animals.

Find one word beginning with each of the letters in the table.
Write a crossword-type clue for each word.

Word	Clue
A	
C	
E	
F	
G	
M	
N	
O	
P	
T	
U	
X	
Y	
Z	

H	C	D	E	M	O	S	O	B	I	R	T	A	S	C
P	H	O	T	O	T	R	O	P	I	S	M	R	U	E
T	R	U	N	S	P	E	C	I	A	L	I	S	E	D
I	O	B	A	S	E	S	S	E	U	N	T	T	L	I
S	M	L	H	A	M	E	L	Y	X	R	O	E	C	A
S	O	E	G	P	P	I	L	I	I	H	C	M	U	P
U	S	H	A	E	N	R	A	G	N	B	H	C	N	A
E	O	E	M	B	R	Y	O	N	I	C	O	E	M	Y
G	M	L	E	E	F	E	T	U	S	Y	N	L	R	A
H	E	I	T	T	A	K	D	O	R	A	D	L	W	D
T	N	X	E	E	T	O	G	Y	Z	E	R	S	E	L
J	Y	A	S	S	E	I	N	A	C	T	I	V	E	O
X	S	E	M	M	S	E	L	L	E	N	A	G	R	O

3 Fill in the empty boxes to show the differences between mitosis and meiosis.

	Meiosis	Mitosis
What does it make?	gametes (sex cells)	
How many new cells does each parent cell make?		2
How many chromosomes are in each new cell?		same as in parent cell
Where does it happen?	in sex organs	
Why does it happen?		so an organism can grow, reproduce and replace damaged cells

B5.1.1–2 What's inside a plant or animal?

Animals and plants are **multicellular organisms**. They are made up of many cells. The cells can be **specialised** to do certain jobs.

In animals and plants, specialised cells of the same type are grouped together to form **tissues**, for example, muscle tissue. Groups of tissues form **organs**, for example, the heart.

B5.1.3–6 What are human stem cells?

Humans develop from a single fertilised egg, or **zygote**. The zygote divides to form new cells. This ball of cells is an **embryo**.

When the embryo consists of eight cells or fewer, all its cells are identical. These are **embryonic stem cells**. Each one could produce any type of cell that the growing human needs.

After the eight-cell stage, most embryo cells become specialised. They form different types of tissue. By about eight weeks, these tissues have grouped together to form the main organs. The growing human is now a **fetus**.

Some cells remain unspecialised until adulthood. These are **adult stem cells**. They can develop into many types of specialised cells.

B5.1.7–9 How do plants grow?

Most plant cells are specialised. Specialised plant cells are grouped together to form tissues such as:

- xylem, which transports water through the plant's organs
- phloem, which transports sugar.

Groups of tissues are organised into organs, for example, leaves, flowers, stems, and roots.

Plant cells at root tips and shoot tips remain unspecialised. So do rings of cells in stems and roots. These are **meristem cells**. They divide to make new cells that can develop into any type of cell.

Meristem cells explain why plants can grow throughout their lives, and why plants can regrow whole new organs if they are damaged.

B5.1.10–11 What are cuttings?

Gardeners sometimes grow plants from **cuttings**. The new plant and the parent plant are genetically identical. They are **clones**.

First, the gardener chooses a plant with features that he wants. He cuts a piece of stem from the plant. This is a cutting. The gardener dips the cut end in **rooting powder**. Rooting powder contains plant hormones, or **auxins**.

H

The meristem cells in the cuttings divide to make new cells. Plant hormones encourage some of these cells to develop into root cells. Other unspecialised cells become tissues that form other organs.

B5.1.12–14 How does light affect plant growth?

Plants tend to grow towards the light. This is **phototropism**. Plants need light to photosynthesise. So phototropism increases a plant's chance of survival.

Phototropism involves **auxins**.

If light is above a growing shoot, auxins spread out evenly. The shoot grows straight up.

light

auxin moves evenly down the stem

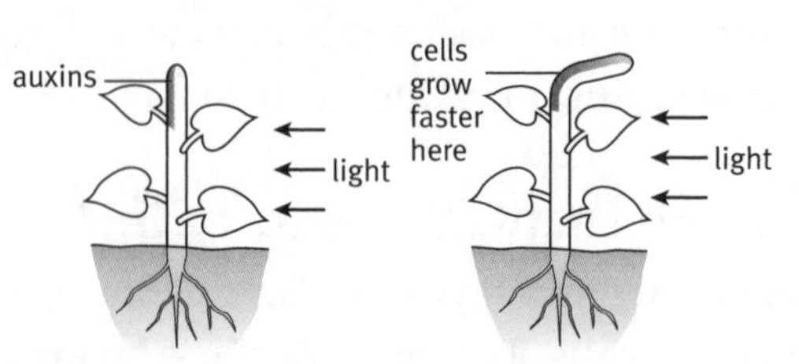

If light comes from the side, auxins move to the shady side. Auxins make cells on the shady side grow faster. So the shoot bends towards the light.

B5.2.1–2 How does mitosis make new body cells?

First, a cell grows. During growth, it makes copies of:

- its **organelles** (specialised parts) including mitochondria
- its chromosomes – these are copied when the two strands of each DNA molecule separate and new strands form alongside them.

Next, **mitosis** happens. In mitosis the chromosome copies separate and go to opposite ends of the cell. Then the whole cell (including the nucleus) divides to make two new cells. The new cells are genetically identical to each other and to the parent cell.

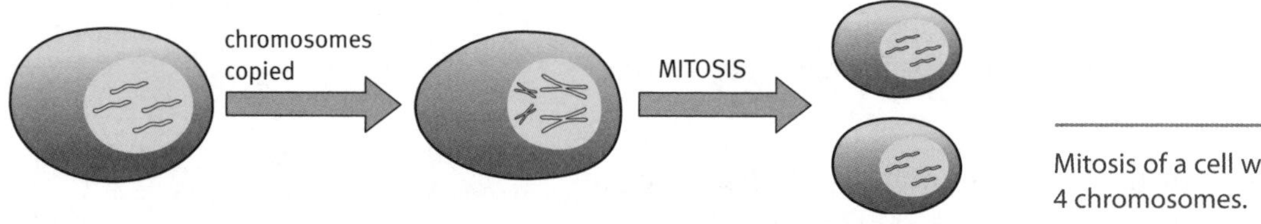

Mitosis of a cell with 4 chromosomes.

B5.2.3–4 How does meiosis make new sex cells?

Meiosis makes sex cells (**gametes**). It happens in sex organs.
In meiosis, a body cell divides twice. The resulting cells may develop into gametes. Gametes are not identical – they each carry different genetic information.

Gametes have half the number of chromosomes as the parent cell.
Human body cells have 46 chromosomes, arranged in 23 pairs.
So human gametes (sperm and egg cells) have only 23 single chromosomes.

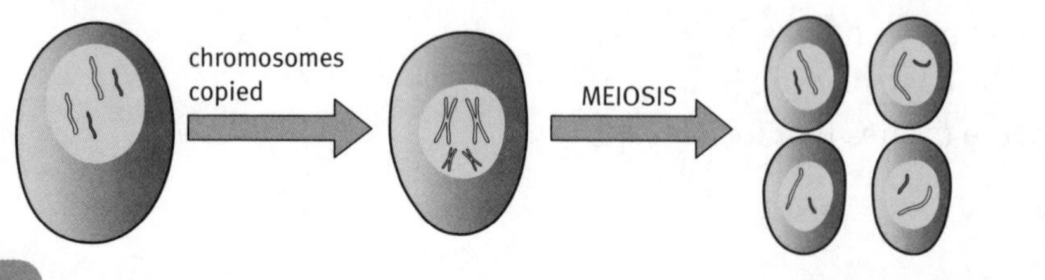

Meiosis of a cell with 4 chromosomes.

Fertilisation

When a human sperm cell fertilises an egg cell, their nuclei join up. The fertilised egg cell (**zygote**) gets one set of chromosomes from each parent. It has 23 chromosome pairs – 46 chromosomes in all.

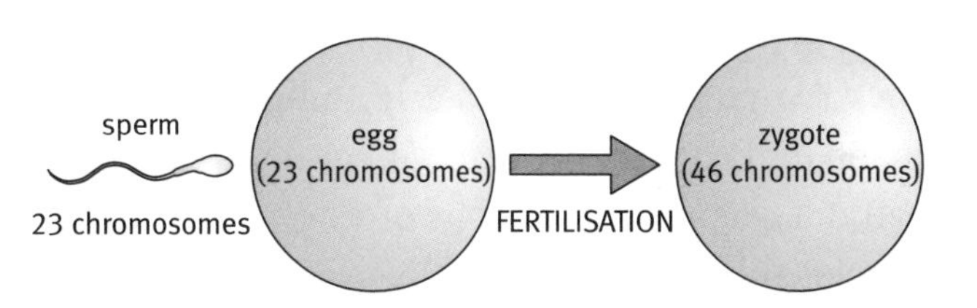

B5.3.1–4 What's in DNA?

The nucleus of each of your body cells contains enough information to determine the characteristics of your whole body. This information is the **genetic code**.

The genetic code is stored in chromosomes. A chromosome is a very long molecule of **DNA** wound around a protein framework. A human DNA molecule is made up of about 30 000 genes.

A DNA molecule contains two strands twisted together in a spiral. This is a **double helix**.

- Each strand is made of four bases: adenine (A), thymine (T), guanine (G), and cytosine (C).
- The bases on the two strands of a DNA molecule always pair up in the same way – A pairs with T, and G pairs with C. This is **base pairing**.

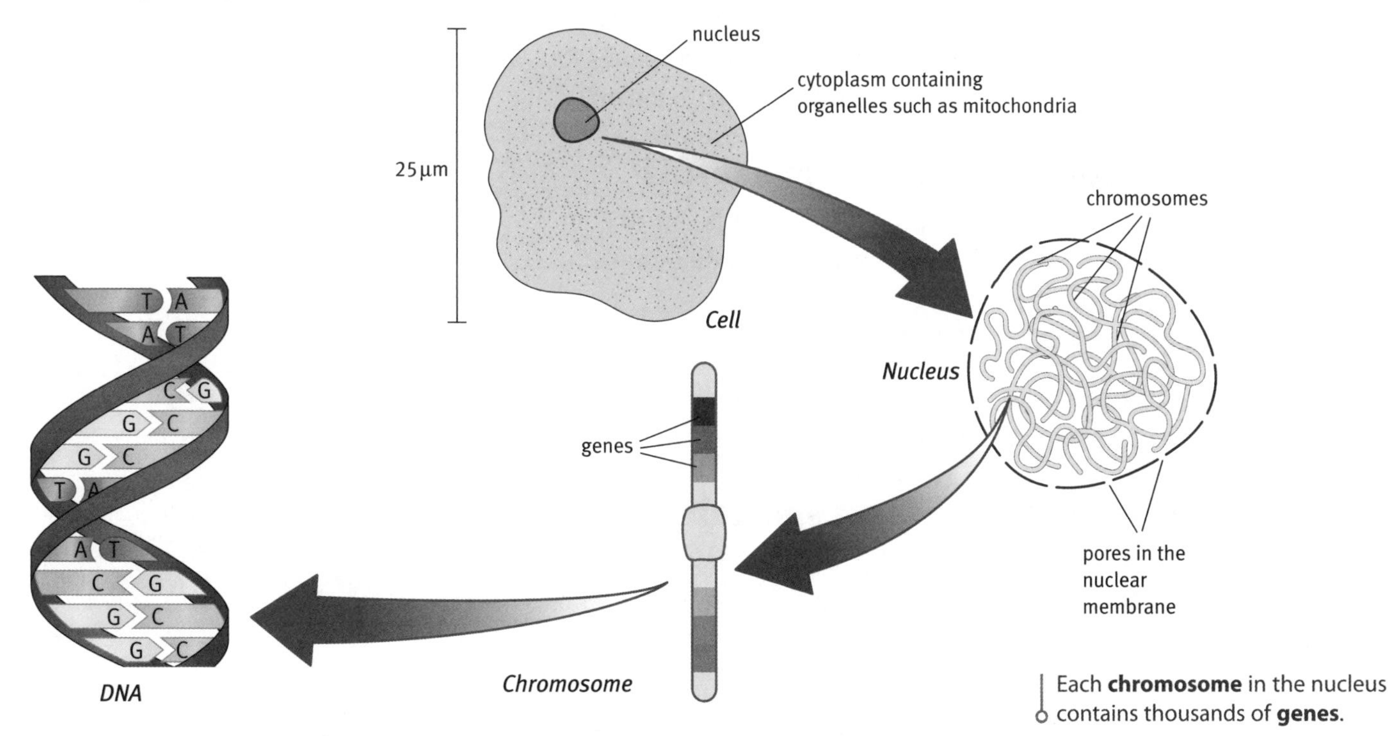

Each **chromosome** in the nucleus contains thousands of **genes**.

Cells make proteins. There are thousands of different proteins. The order of the bases in a gene is the genetic code for making a particular protein.

H Cells make proteins by joining together **amino acids**. Each protein has a certain combination of amino acids joined together in a particular order. The order of bases in a gene is the code for building up amino acids in the correct order to make a particular protein.

B5.3.5–6 Where do cells make proteins?

DNA contains the genetic code for making proteins. DNA is in the nucleus of a cell. But new proteins are made in the cell cytoplasm. Genes do not leave the cell nucleus. Instead, a copy of the gene is made. This carries the genetic code to the cytoplasm.

H The copy of the gene that carries the code is called **messenger RNA**.

unspecialised cell

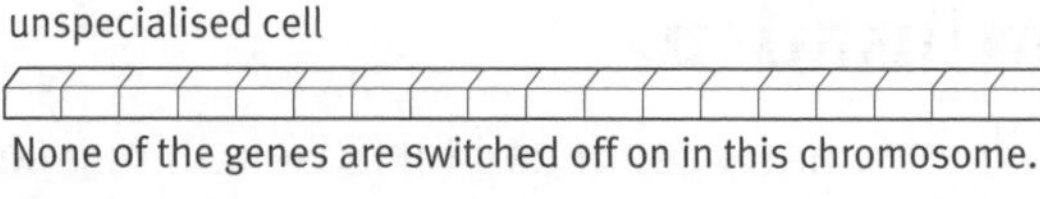

None of the genes are switched off on in this chromosome.

hair cell

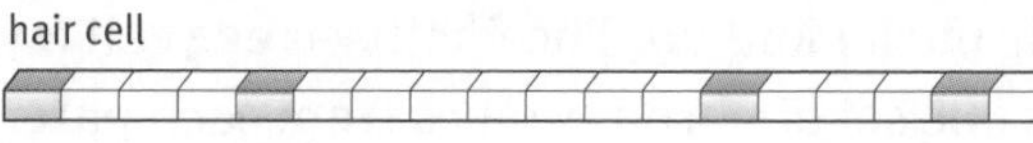

salivary gland cell

Different genes are switched off in specialised cells.

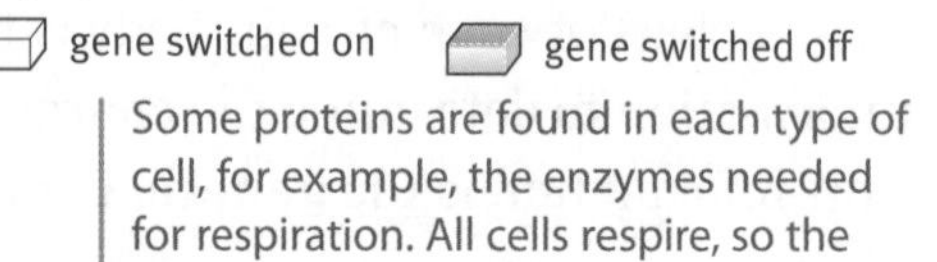

Some proteins are found in each type of cell, for example, the enzymes needed for respiration. All cells respire, so the genes needed for respiration are switched on in all cells.

B5.3.7–8 Why are some genes switched off?

The nucleus of each of your body cells contains an exact copy of the DNA of the original zygote. So every cell contains the same genes. But not all these genes are active in every cell. Each cell makes only the proteins it needs to be a particular type of cell. Genes that give instructions to make other proteins are not active; they are **switched off**.

For example, hair cells make keratin. In hair cells, the genes for the enzymes that make keratin are switched on. Hair cells do not make muscle. So the genes for the enzymes that make muscle are switched off.

Salivary gland cells make amylase. In these cells, the genes for the enzymes that make amylase are switched on. The genes for the enzymes that make keratin and muscle are switched off.

B5.3.9–10 How are stem cells useful?

Scientists are doing a lot of research on stem cells. Adult stem cells, and embryonic stem cells, may make specialised cells to:

- replace damaged tissue
- treat some diseases.

There are ethical issues to consider when using embryonic stem cells, so there are government regulations to control this research.

B5.3.11 How do clones form specialised cells?

H Scientists have made clones of animals such as sheep, and of tissues and organs of other mammals, including humans. The process involves reactivating, or switching on, inactive genes in adult body cell nuclei.

This is how to produce an organ or tissue needed by a patient:

- Take a nucleus out of a human egg cell. Replace it with a nucleus from a body cell of the patient.
- The egg cell divides by mitosis and makes an embryo. The genes in the embryo are the same as the patient's genes.
- After 5 days, put stem cells from the embryo in a dish of nutrients.
- The stem cells can develop into different types of tissue or organ.
- Transplant the organ or tissue required into the patient.

The nucleus is taken out of a human egg cell. It is replaced with a nucleus from a body cell of the patient.

The egg cell is triggered to develop into an embryo.

3 days

pipette

5 days
Cells are removed from the embryo with a pipette.

stem cells

Stem cells from the embryo are grown in a dish containing nutrients.

Stem cells develop into different tissues and organs. These can be used for medical treatment.

Therapeutic cloning.

Use extra paper to answer these questions if you need to.

1 Write **P** next to the statements that are true for plants only, **A** next to the statements that are true for animals only, and **B** next to the statements that are true for both.

- **a** Can grow in height and width for their whole lives. __
- **b** Contain organs. __
- **c** Grow only at meristems. __
- **d** Contain groups of similar cells called tissues. __
- **e** Contain specialised cells. __
- **f** Can regrow whole organs if they are damaged. __

2 Use words from the box to fill in the gaps. Each word may be used once, more than once, or not at all.

mitochondria **mitosis** **organelles** **DNA** **cell** **strands** **meiosis** **chromosomes** **nucleus**

In the cell cycle, cells grow. During cell growth, the numbers of __________ increase, including __________. Also, the __________ are copied. This happens when two strands of each __________ molecule separate and new __________ form alongside them.
Then __________ happens. During this process, copies of the __________ separate and the __________ divides.

3 Fill in the empty boxes.

Species	Number of chromosomes in sex cell	Number of chromosomes in body cell
horse	32	
wolf		78
carp	52	
		46

4 Draw lines to match each base in DNA with the base it pairs up with.

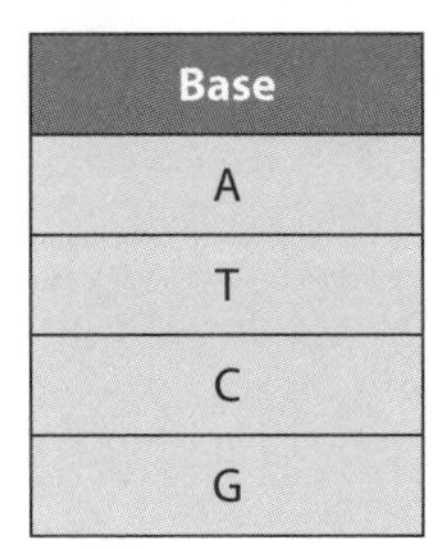

Base		Base
A		A
T		T
C		C
G		G

5 Write **T** next to the statements that are **true**. Write corrected versions of the statements that are **false**.

- **a** Up to the 16-cell stage, all the cells in a human embryo are identical.
- **b** Embryonic stem cells can produce any type of cell needed by a human.
- **c** Adult stem cells can produce any type of cell needed by a human.
- **d** A zygote divides by meiosis to form an embryo.
- **e** A zygote contains a set of chromosomes from each parent.
- **f** In a 16-cell human embryo, most of the embryo cells are specialised.
- **g** Groups of specialised cells in an animal are called zygotes.
- **h** Cell division by meiosis produces new cells that are genetically identical to the parent cell.
- **i** Cell division by mitosis produces new cells that are genetically identical to each other.
- **j** Meiosis makes gametes.

6 Tick the boxes to show which genes are switched on in the cells below. Hint: nails and hair contain a protein called keratin.

Cell	Are these genes switched on?		
	gene to make keratin	gene to make salivary amylase	gene for respiration
nail			
hair			
embryonic stem cell			
salivary gland cell			
muscle cell			

7 Write down where in a cell the processes below happen.

- **a** A copy of a gene is made.
- **b** Proteins are produced.
- **c** Aerobic respiration.

H **8** Draw arrows to show where the light is coming from for each of the plant shoots below. Shade each plant shoot to show where the concentration of auxins is greatest.

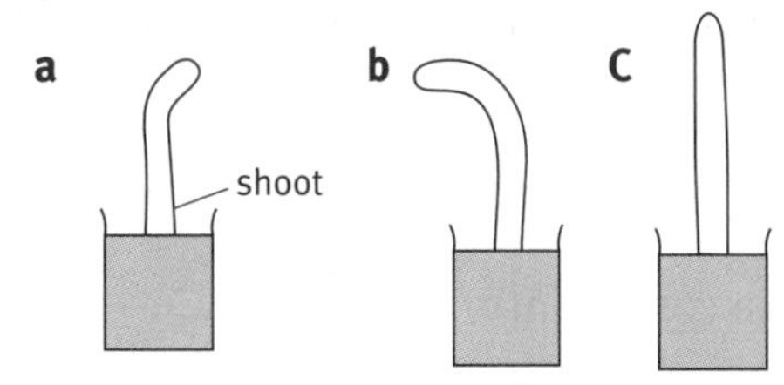

9 Name the chemical that carries the genetic code from the nucleus to the cytoplasm of a cell.

10 Explain how DNA codes for the making of proteins.

1 Giraffes have 30 chromosomes in each body cell.

a Finish the sentences by choosing the best words and numbers from the box.

Use each word or number once, more than once, or not at all.

gametes	zygotes	different	ovaries	testes
penis	identical	1 2 4	15 30	60

Giraffes make sex cells by meiosis. Sex cells are also called ______________. In male giraffes, meiosis happens in the ______________. In meiosis, one body cell divides to make ________ sex cells. Each of these cells carries ______________ genetic information. There are ________ chromosomes in one giraffe sex cell. [5]

b After sexual intercourse, the nucleus of a male sex cell joins to the nucleus of a female sex cell.

i Give the name of the female sex cell.

______________________________ [1]

ii Give the name of the process in which the nucleus of a male sex cell joins to the nucleus of a female sex cell.

______________________________ [1]

c The steps below describe how body cells grow and divide in giraffe embryos.

They are in the wrong order.

A The chromosome copies separate and go to opposite ends of the cell.

B These are identical to each other and to the parent cell.

C The cell makes copies of its specialised parts, including the chromosomes.

D The cell divides to make two new cells.

Fill in the boxes to show the correct order. [3]

☐ ☐ ☐ ☐

Total [10]

2 James has a rose plant.

He cuts a piece of stem from the plant. This is a cutting.

He dips the end of the cutting in rooting powder, and plants it in compost.

The cutting grows roots.

A new rose plant grows.

a **i** Why do gardeners take cuttings?

Tick the **two best** reasons.

They can grow many new plants quickly and cheaply. ☐

They can grow many new plants with different features by taking cuttings from just one plant. ☐

Plants grown from cuttings are more resistant to disease than plants grown from seed. ☐

They can reproduce a plant with exactly the features they want. ☐

Plants grown from cuttings are stronger than plants grown from seed ☐ [1]

H

ii Give the name of the plant hormone in rooting powder.

________________ [1]

b Vijay is investigating cuttings.

He takes 10 stems from a hedge plant.

He removes the leaves from the bottom 5 cm of the stems.

He dips the cut ends in water.

He dips the wet ends of 5 stems in rooting powder.

He plants all 10 stems in compost in plant pots.

After two weeks, he removes the stems from the pots.

He washes the compost off the stems.

He counts the roots that are growing in each stem.

He writes his results in a table.

	Number of roots after 2 weeks					
	Test 1	Test 2	Test 3	Test 4	Test 5	Mean
Cuttings with rooting powder	7	6	9	5	5	
Cuttings without rooting powder	3	4	0	2	0	

i Complete the table by calculating the mean values. Record the mean values in the empty boxes, rounding each answer to the nearest whole number. [2]

ii Suggest why Vijay used 5 cuttings with rooting powder, and 5 cuttings without, instead of using just one cutting in each of the two conditions.

____

______________________________ [1]

iii Write a conclusion to Vijay's experiment.

____

______________________________ [1]

c Vijay finds the graph below in a text book. Describe what the graph shows.

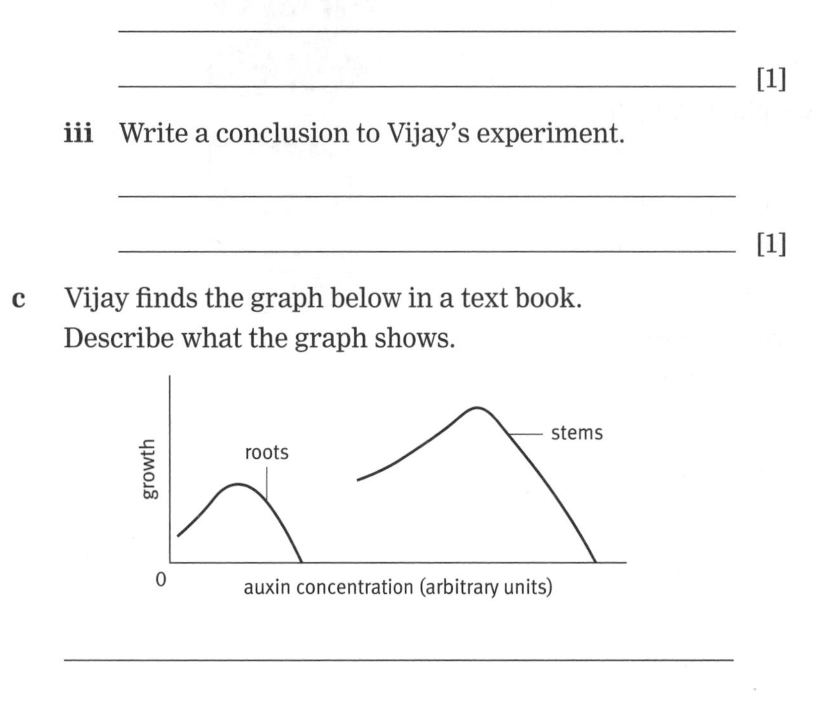

____

____

______________________________ [3]

Total: [9]

Exam tip

When writing conclusions, make sure you clearly describe any correlations between factors.

Exam tip

When describing what a graph shows, make sure you describe relationships between the factors of the axes for each part of the graph.

3 A company sells 'stem cell gift certificates'.

When a baby is born, a nurse takes blood from the baby's umbilical cord. The company separates stem cells from the blood. It stores the stem cells at –180 °C for 25 years.

a i What are stem cells?

Put a tick next to the **best** definition.

Stem cells are unspecialised cells. They join together to make specialised cells. ☐

Stem cells are unspecialised cells. They divide and develop into specialised cells. ☐

Stem cells are specialised cells. They divide and develop into unspecialised cells. ☐

Stem cells are specialised cells. They join together to make unspecialised cells. ☐ [1]

ii The company says that doctors can use the stem cells to treat illnesses the baby may get in future.

How might doctors use the stem cells to treat heart disease?

Put a tick next to the **most likely** answer.

- They will make a heart disease vaccine from the stem cells. ☐
- They will grow heart muscle cells from the stem cells. ☐
- They will make a heart disease medicine from the stem cells. ☐
- They will inject stem cells into the patient's bloodstream. ☐ [1]

iii Some friends are discussing removing and storing umbilical cord blood.

Here are their ideas.

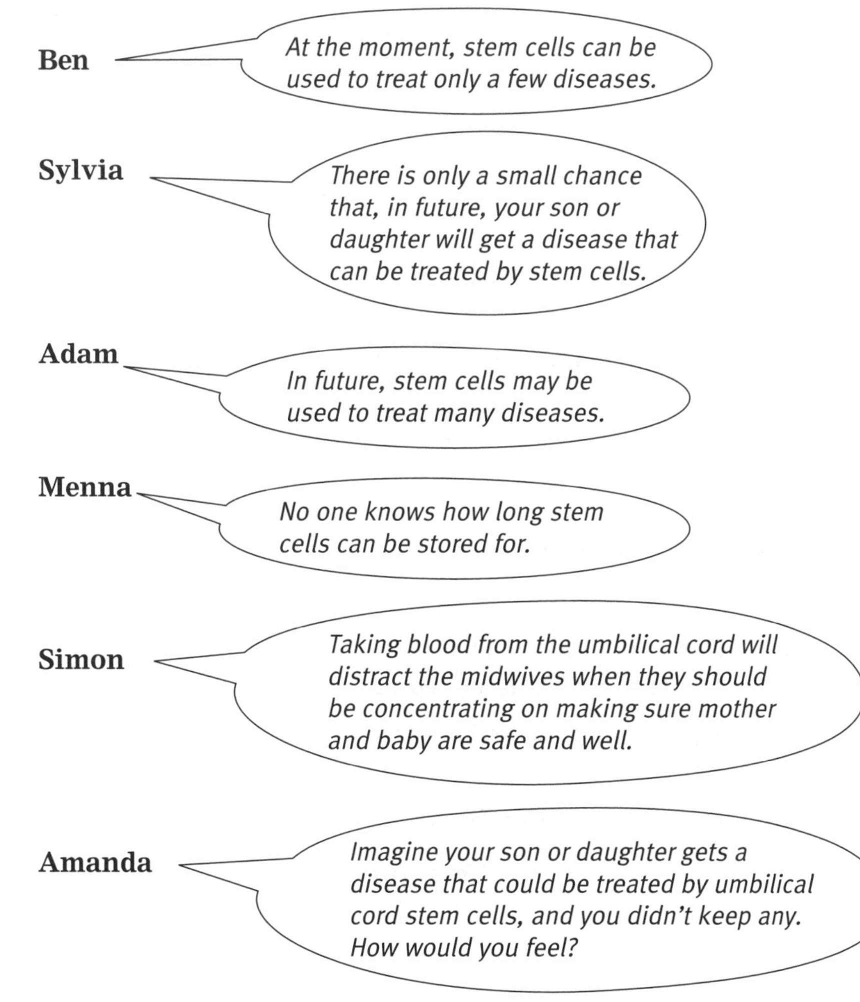

Write a paragraph to advise a pregnant woman whether she **should *or* should not** have her baby's umbilical cord blood stored.

Select reasons from the speech bubbles to support your answer.

____

____

____

____ [3]

b One source of stem cells is umbilical cord blood.

i Name **one other** source of stem cells.

____ [1]

ii Give one **problem** of using stem cells from this source to treat disease or replace damaged tissues.

____

____ [1]

Total: [7]

4 Describe the differences and similarities between mitosis and meiosis.

✎ The quality of written communication will be assessed in your answer to this question.

Write your answer on separate paper or in your exercise book.

Total [6]

Going for the highest grades

H **5** Explain why, in a specialised body cell, many genes are inactive (switched off).

Describe the conditions in which inactive genes in a body cell may be reactivated (switched on) to form cells of all tissue types.

✎ The quality of written communication will be assessed in your answer to this question.

Write your answer on separate paper or in your exercise book.

Total [6]

1 Match each word with its definition.

Word
behaviour
stimulus
response

Definition
a change in the environment
an action caused by a change in the environment
anything an animal does

2 The diagram shows a reflex arc. Use these words to label the diagram.

receptor **sensory neuron** **spinal cord**
motor neuron **effector**

Exam tip

Make sure you can label a reflex arc – they often come up in exam questions.

ouch!

muscle cell – an ________________

skin cell – a ________________

3 The diagram shows a motor neuron.
Complete the labels to describe what each part of the cell does.

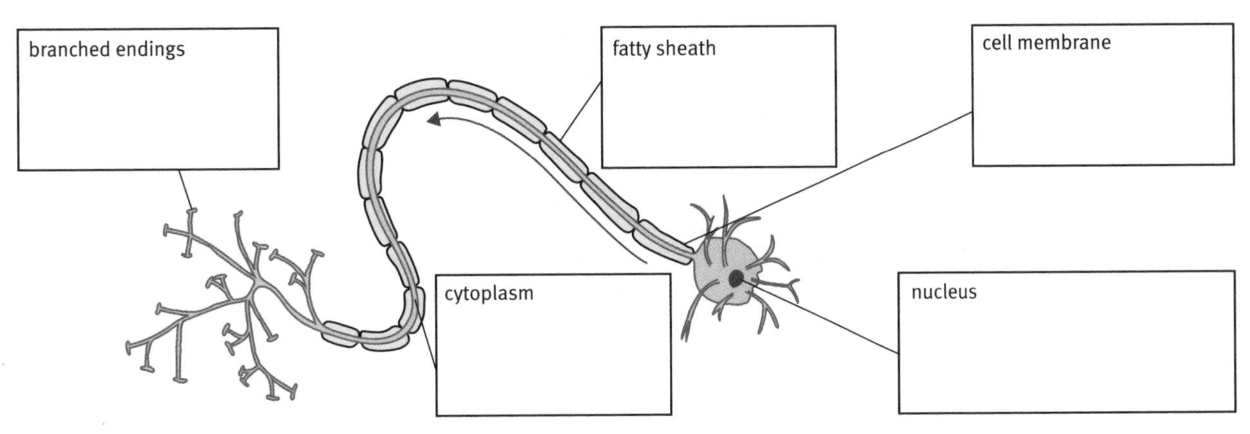

B 6

4 The diagrams show how a nerve impulse crosses a synapse.
They are in the wrong order.

- Number each box to show the correct sequence.
- Write notes next to each diagram to explain the process.

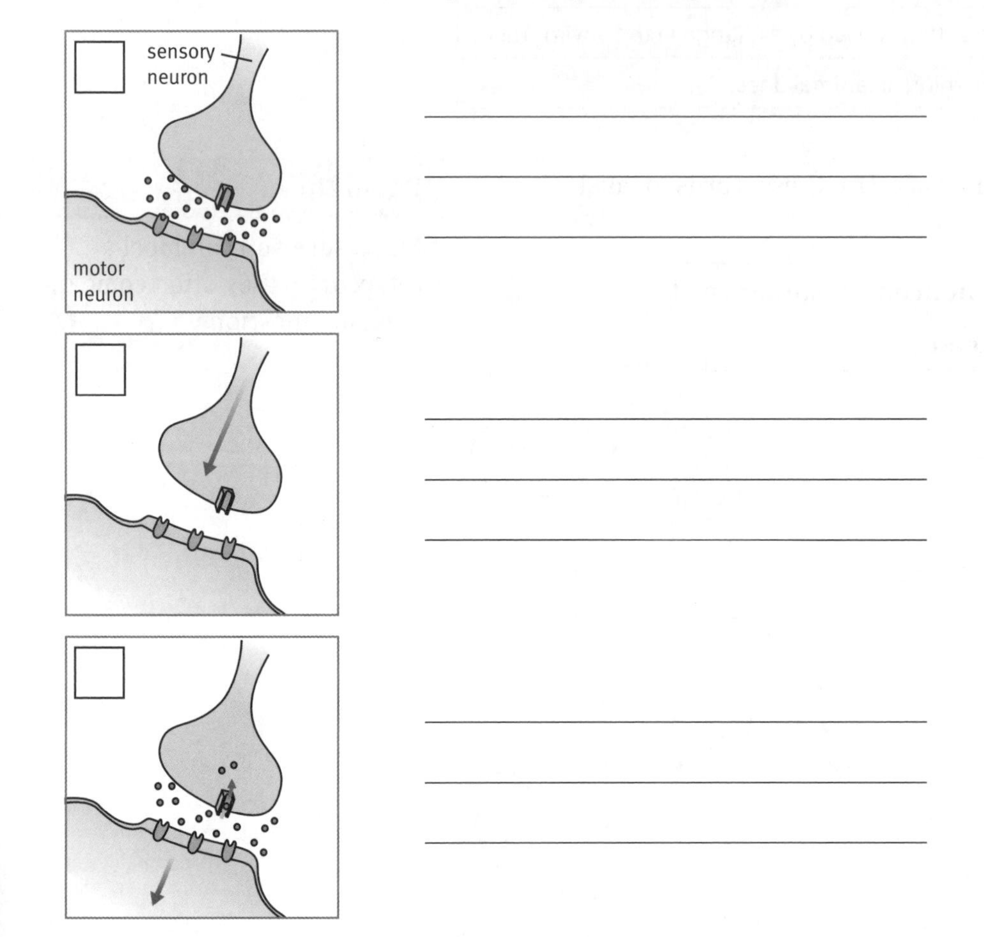

5 The cartoon shows the benefits of one conditioned reflex action.
Complete the thought bubbles to describe the benefits.

B6.1.1 How do animals respond to changes?

You are at a party. The room gets hotter. You sweat more.
A change in your environment, or **stimulus** (temperature increase), has triggered a **response** (more sweating).

B6.1.2–5 What are simple reflex actions?

Simple reflex actions happen quickly. They are **involuntary** (automatic).

Simple animals rely on reflex actions for most of their behaviour. So they always respond in the same way to a particular stimulus. Woodlice move away from light, for example.

Reflex actions help simple animals respond to stimuli in ways that help them survive, including finding food and hiding from predators.

A human baby shows **newborn reflexes**, including:
- **grasping** – tightly gripping a finger in her palm
- **sucking** a nipple or finger in her mouth
- **stepping** when her feet touch a flat surface.

Adults show reflex actions, too. For example:
- The **pupil reflex** stops light damaging cells in your retina. In bright light, some of the muscles in your iris contract. Your pupil gets smaller, so less light enters your eye.
- If someone hits your leg just below the knee, your thigh muscle contracts and your leg straightens. This is the **knee jerk reflex**.
- If you pick up something hot, you drop it before you feel it.

B6.1.6–10 How do we respond to stimuli?

To respond to stimuli, an animal needs:
- **Receptors** to detect stimuli. These include:
  - single cells, such as pain sensor cells in the skin
  - cells in complex organs, for example, cells in the retina.
- **Processing centres** to receive information and co-ordinate responses.
- **Effectors** to respond to stimuli. These include:
  - **muscle cells**, which contract to move a part of the body
  - **glands**, whose cells release chemical **hormones**.

As multicellular organisms evolved, their bodies developed complex nervous and hormonal communication systems.
- **Nerve impulses** are electrical signals. They bring about fast, short-lived responses.
- **Hormones** travel in the blood. They cause slower, longer-lasting responses, for example:
  - The pancreas makes **insulin** to control blood sugar.
  - **Oestrogen** helps to control the female menstrual cycle.

B6.2.1–10 How does the nervous system work?

In the nervous system, **neurons** (nerve cells) link receptor cells to effector cells. Neurons transmit electrical impulses when stimulated.

A neuron has an **axon**. This is a long extension of the cytoplasm. It is surrounded by a cell membrane. Some axons are surrounded by a **fatty sheath**. This insulates the neuron from nearby cells, so electrical signals can pass along it quickly.

In humans and other vertebrates, the **central nervous system** (CNS) is made up of the spinal cord and brain. The CNS coordinates an animal's responses to stimuli.

The **peripheral nervous system** (PNS) links the CNS to the rest of the body. It is made up of sensory and motor neurons.

In a simple reflex, impulses move from one part of the nervous system to the next in a **reflex arc**.

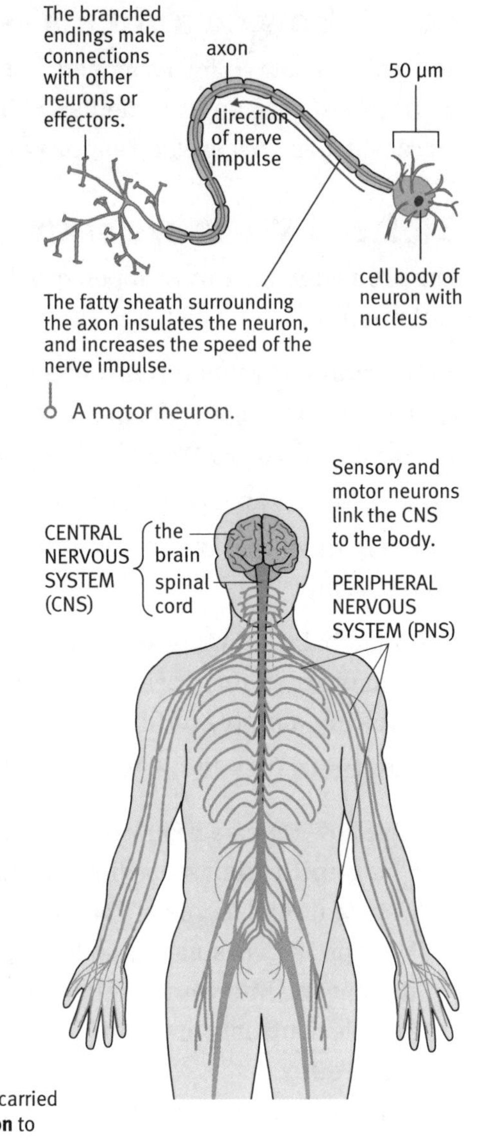

A motor neuron.

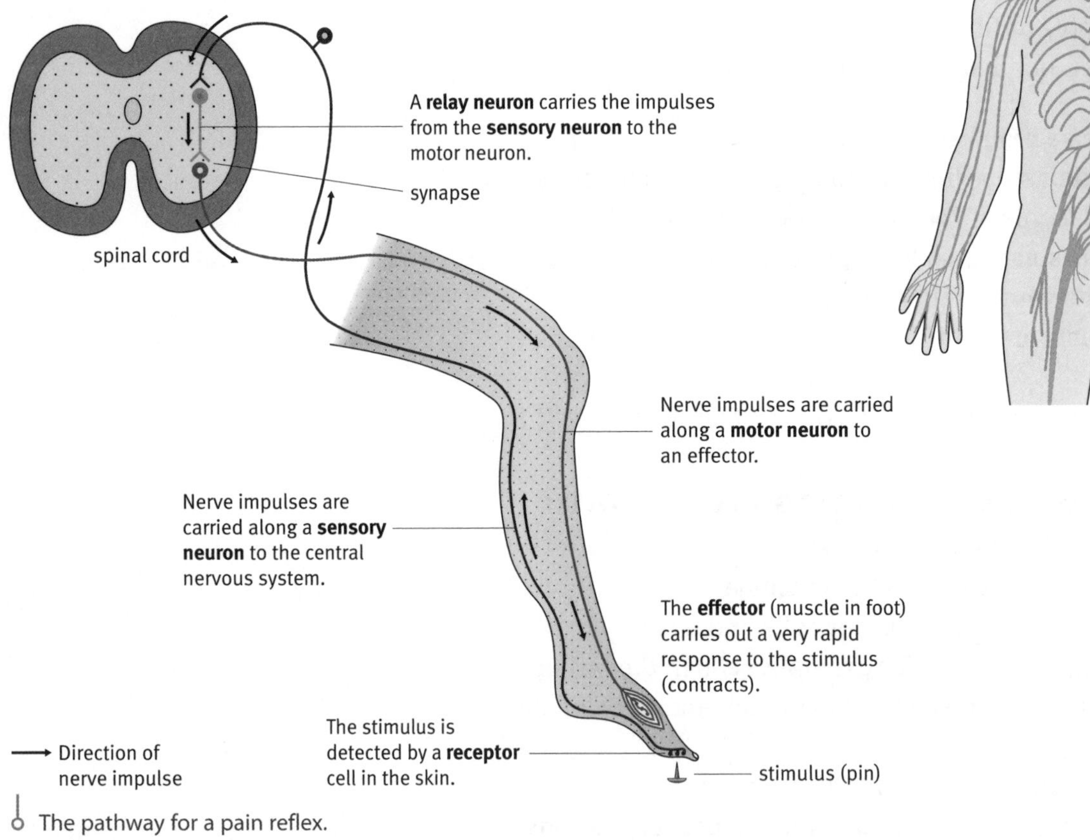

The pathway for a pain reflex.

In a reflex arc, the neurons are arranged in a fixed pathway. Information does not need to be processed, so responses are automatic. This makes reflex responses very quick.

B6.2.11–13 What are synapses?

There are tiny gaps (**synapses**) between neurons in a reflex arc. Nerve impulses must cross these gaps to get from one neuron to the next. Special chemicals pass impulses across synapses.

These chemicals are called **transmitter substances**. The diagrams show how they work.

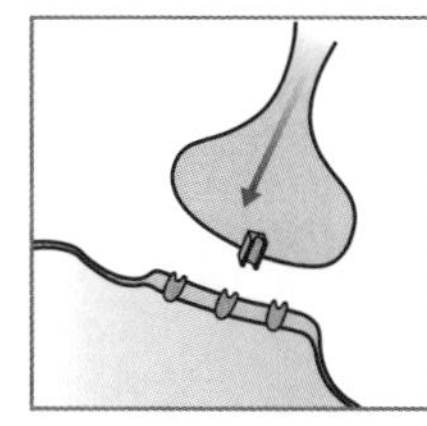

1 A nerve impulse gets to the end of a sensory neuron.

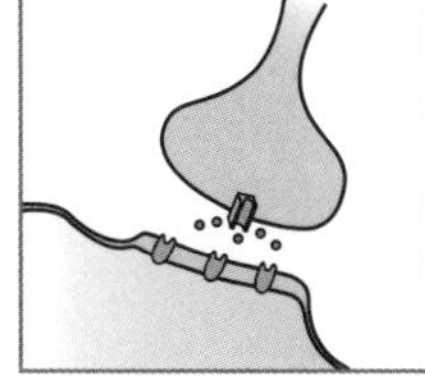

2 The sensory neuron releases the transmitter chemical into the synapse.

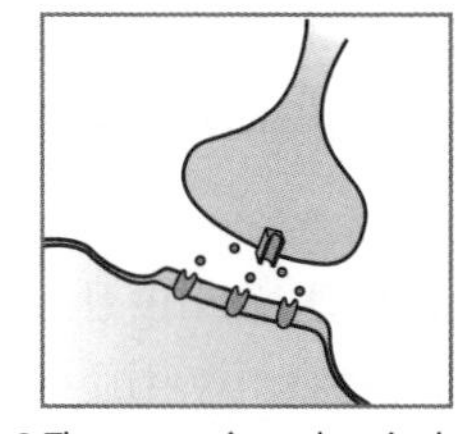

3 The transmitter chemical diffuses across the synapse.

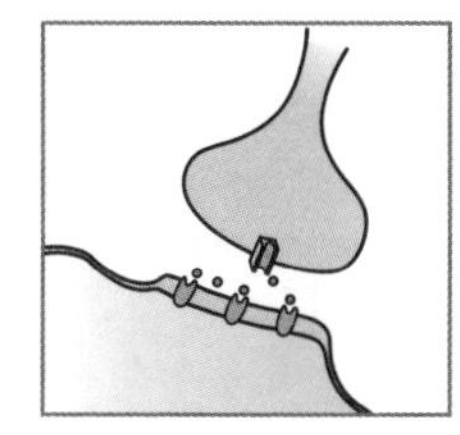

4 The transmitter chemical arrives at receptor molecules on the motor neuron's membrane. Its molecules are the correct shape to bind to the receptor molecules.

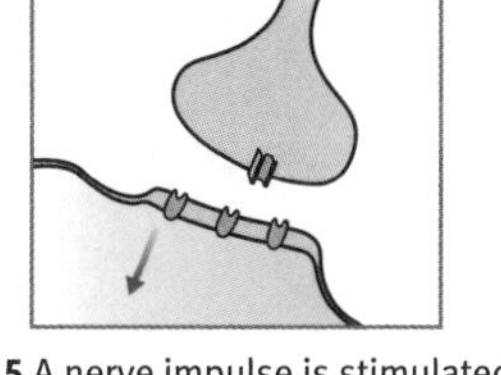

5 A nerve impulse is stimulated in the motor neuron.

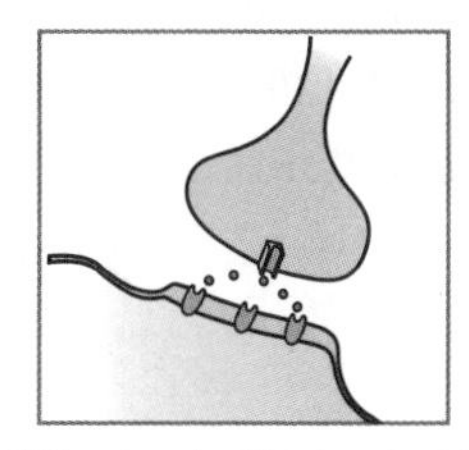

6 The chemical is absorbed back into the sensory neuron to be used again.

B6.2.14–16 How do drugs affect synapses?

Some drugs, including Ecstasy, beta blockers, and Prozac, make it more difficult for impulses to get across synapses.

Serotonin (a transmitter chemical) is released at some brain synapses. This gives a feeling of pleasure. Sensory neurons later remove the serotonin.

Ecstasy (MDMA) blocks the places that remove serotonin. So the serotonin concentration in the synapse increases. This may make Ecstasy users feel happy for a while. But Ecstasy is harmful.

B6.2.17–18 What's inside the brain?

Different regions of the brain do different jobs. Neuroscientists map the brain by:

- studying patients with brain damage
- electrically stimulating different parts of the brain
- doing MRI scans.

The **cerebral cortex** is the part of the brain most closely linked to intelligence, language, memory, and **consciousness** (being aware of yourself and your surroundings).

B6.3.1–4 Can you learn reflex responses?

Animals can learn a reflex response to a new stimulus. The new stimulus (the **secondary stimulus**) becomes linked to the primary stimulus. This is **conditioning**.

Pavlov taught a dog to salivate when it heard a bell ring:

- The dog's simple reflex was to salivate when it was given food.
- Pavlov rang a bell while the dog was eating.
- After a while, the dog salivated every time it heard the bell, even if there was no food.

H In this **conditioned reflex**, the stimulus (hearing the bell) became linked to food. The final response (salivating) had no direct connection to the stimulus.

Conditioned reflexes are a simple form of learning. They increase an animal's chance of survival. For example, many bitter-tasting caterpillars are brightly coloured. A bird tastes these caterpillars, and learns not to eat them. This helps the caterpillars to survive.

B6.3.5 Can you control reflexes?

Sometimes your brain consciously changes a reflex response.

- You pick up a hot object. An impulse travels to your spinal cord. Your reflex response is to drop the plate.
- But another nerve impulse travels up your spinal cord to your brain. The impulse returns down a motor neuron. This makes a muscle movement in your arm that stops the reflex response.
- You keep hold of the hot object until you can put it down safely.

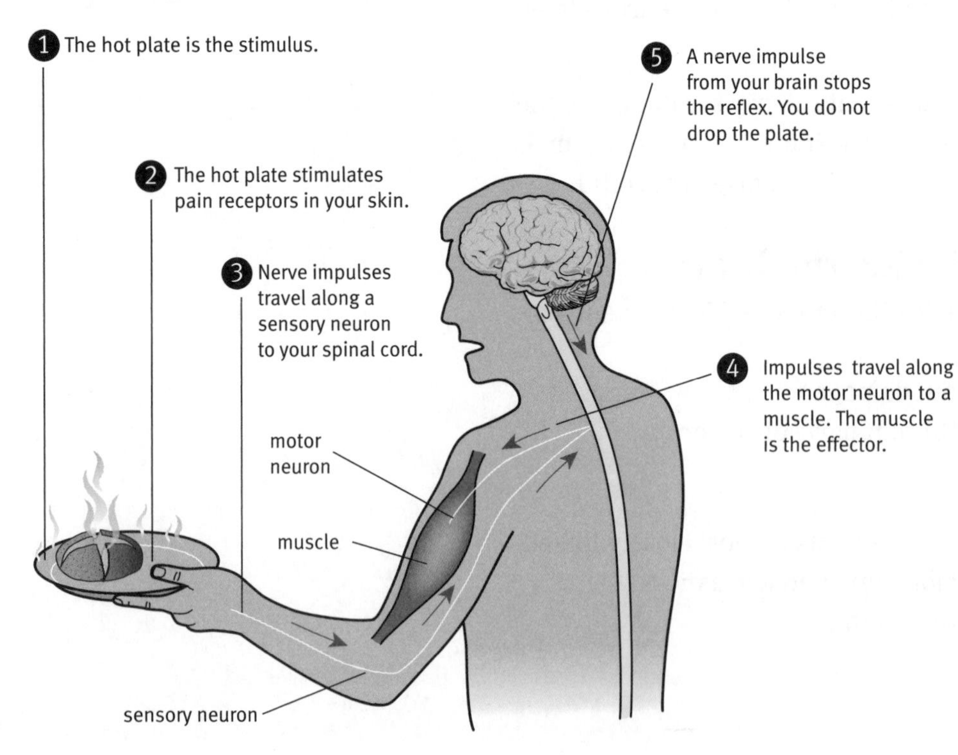

6.4.1–7 How do we develop complex behaviour?

Mammals can change their behaviour as a result of new experiences. This is **learning**.

Mammal brains have billions of neurons, connected in **pathways**. Learning creates new pathways. Young animals quickly create many new pathways. Adults also make new pathways. Here's how:

- You experience something new.
- A nerve impulse travels along a particular pathway, from one neuron to another, for the first time. This makes new connections between the neurons.
- You repeat the experience.
- More impulses go along the same pathway. The connections get stronger.
- Nerve impulses travel along the pathway more easily. It is easier to respond in the way that you practised.

H The variety of possible brain pathways means that animals can adapt to new situations. They have a better chance of survival.

Early humans evolved large brains, which increased their chance of survival.

H Evidence suggests that children can only acquire some skills at a certain age. For example, a **feral** (wild) child cannot learn to speak if they are found after the best age for learning language skills.

B6.4.8–12 What is memory?

Memory is the storage and retrieval of information by the brain.

- **Short-term memory** lasts about 30 seconds. Scientists see this as an active **working memory**, where you can hold and process information you are thinking about now.
- **Long-term memory** is a seemingly limitless store of information that can last a lifetime.
- **Sensory memory** stores sound and visual information.

Psychologists have developed **models of memory**. All the models have limitations.

- **Multistore model** – this explains how some information is passed to the long-term memory store and how some information is lost.
- **Working memory** – you are more likely to remember something if you process it deeply. This happens if you can find patterns, or if you can organise the information.
- **Repetition** – psychologists think that repeating information moves information from the short-term to long-term memory.

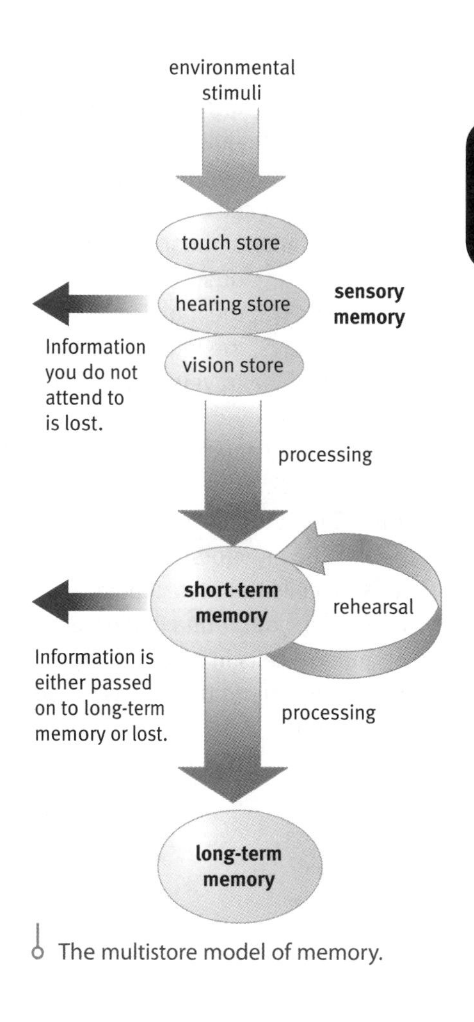

The multistore model of memory.

B 6

B6: Brain and mind

Use extra paper to answer these questions if you need to.

1 Write **S** next to the sentences that best apply to simple animals, like woodlice. Write **C** next to the sentences that best apply to complex animals, like horses.

- **a** These animals rely on reflex actions for most of their behaviour. __
- **b** These animals can change their behaviour. __
- **c** The animals find it difficult to respond to new situations. __
- **d** These animals can learn to link a new stimulus to a reflex action. __

2 Give the names of:

- **a** the two parts of the central nervous system
- **b** two types of neurons in the peripheral nervous system.

3 Draw lines to match each 'message carrier' in the left column to two characteristics in the right column.

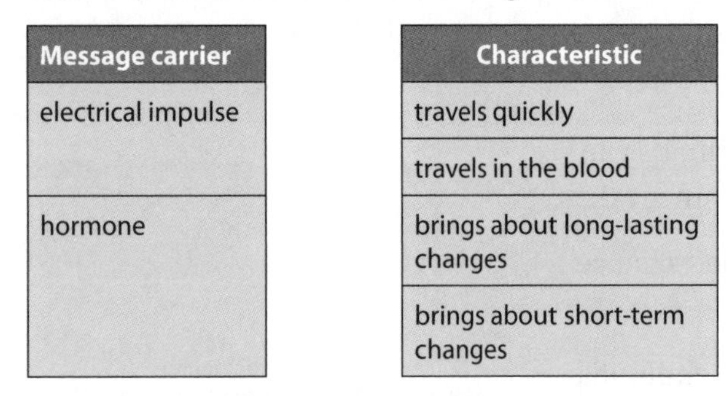

Message carrier
electrical impulse
hormone

Characteristic
travels quickly
travels in the blood
brings about long-lasting changes
brings about short-term changes

4 The stages below describe how information is passed along a simple reflex arc. They are in the wrong order. Write the letters of the stages in the correct order.

- **A** A receptor cell detects dust in your eye.
- **B** The effectors (muscles in your eyelid) blink to remove the dust.
- **C** In the CNS, the impulse passes to a relay neuron.
- **D** In the CNS, the impulse passes to a motor neuron.
- **E** An impulse travels along a motor neuron to the effectors.
- **F** A sensory neuron carries electrical impulses to your CNS.

5 Write **S** next to the sensors in the list below, and **E** next to the effectors.

- **a** skin cells that detect pain __
- **b** cells in your retina that detect light __
- **c** muscle cells in a baby's finger __
- **d** a sweat gland that releases sweat when you are nervous __
- **e** taste buds on your tongue __
- **f** semi-circular canals in your ear that detect movement __
- **g** the salivary gland that releases saliva when you smell food __

6 Complete the sentences below.

- **a** Early humans had a better chance of survival because they evolved big b_______.
- **b** Your brain has billions of n_______.
- **c** The cerebral cortex is the part of your brain most concerned with c_______, l_______, i_______ and m_______.
- **d** Different areas of the cortex have different jobs. Scientists have mapped these areas by...

7 The stages below describe how new pathways develop in your brain when you learn to iceskate. They are in the wrong order. Write the letters of the stages in the correct order.

- **A** You ice skate for the first time.
- **B** More impulses go along the same pathway.
- **C** There is now a new neuron pathway.
- **D** You go ice skating again.
- **E** An impulse travels along a certain pathway, passing from one neuron to another, for the first time.
- **F** The connection between the neurons gets stronger.
- **G** You now find it much easier to iceskate.

8 Write **T** next to the statements that are **true**. Write corrected versions of the statements that are **false**.

- **a** Short-term memory is a seemingly limitless store of information.
- **b** You are more likely to remember information if you can find patterns in the information.
- **c** Repetition moves information from your long-term memory to your short-term memory.
- **d** The multistore model of memory states that any information you ignore is lost from your memory.
- **e** Your sensory memory stores only memories linked to sounds.
- **f** Short-term memory lasts about 30 seconds.

H **9** Explain why:

- **a** reflex responses are automatic and very quick
- **b** ecstasy increases the concentration of serotonin in synapses
- **c** conditioned reflexes increase an animal's chance of survival
- **d** mammals can adapt to new situations.

1 This question is about reflex actions.

a Sarah is five months old. Look at the list of things that she does.

Tick the **two** actions that are newborn reflexes.

She grips a finger that is put into the palm of her hand. ☐

She stops crying when her sister sings to her. ☐

She steps when her feet touch a flat surface. ☐

She cries when her favourite toy is taken away. ☐

She goes to sleep in her pram. ☐ [2]

b Until she was two months old, Sarah sucked anything that was put into her mouth.

Complete the sentences.

Sucking is the response to the stimulus of ____________

____

The newborn sucking reflex helped Sarah's survival by making sure she got enough ____________. [2]

c Humans rely on reflex actions for only some of their behaviour.

Worms rely on reflex actions for most of their behaviour.

Give one advantage to worms of relying on simple reflex actions for most behaviour.

____

______________________________ [1]

Total [5]

2 Describe and explain three methods that help humans to remember information.

✎ The quality of written communication will be assessed in your answer to this question.

Write your answer on separate paper or in your exercise book.

Total [6]

3 Tom and Sam investigate the effect of caffeine on reaction time.

Tom holds a metre ruler. Sam watches him.

Tom drops the ruler.

He measures the distance the ruler falls before Sam catches it.

He does the test five times, and records the results in a table.

B6

Next, Sam drinks a caffeine drink, and Tom repeats the tests.

The results are in the table opposite.

The shorter the distance the ruler falls before Sam catches it, the quicker Sam's reaction time.

	Distance ruler falls before Sam catches it (cm)					
	Test 1	Test 2	Test 3	Test 4	Test 5	Mean
before caffeine drink	51	45	50	47	48	
after caffeine drink	30	33	32	29	35	

a **i** Complete the table by calculating the mean values for the distance the ruler falls before Sam catches it, both before and after the caffeine drink.

Record the mean values in the empty boxes, rounding each answer to the nearest whole number. [2]

ii Comment on what the investigation tells you about the effect of caffeine on reaction time.

____

______________________________ [1]

iii Explain why a person's reaction cannot be instantaneous.

____

______________________________ [2]

b Pippa and Kate investigate the effect of another factor on reaction time.

Their results are in the table.

	Distance ruler falls before Kate catches it (cm)					
	Test 1	Test 2	Test 3	Test 4	Test 5	Mean
in a silent room	60	62	58	59	61	60
with loud music playing	75	77	75	73	75	75

i Write a conclusion for Pippa and Kate's investigation, saying what the investigation shows.

____

______________________________ [2]

ii Pippa wants to investigate the effect of alcohol on Kate's reaction time.

Describe an ethical argument **against** investigating this factor.

____

______________________________ [1]

Total [8]

4 a Draw a line to match each part of the nervous system to its job.

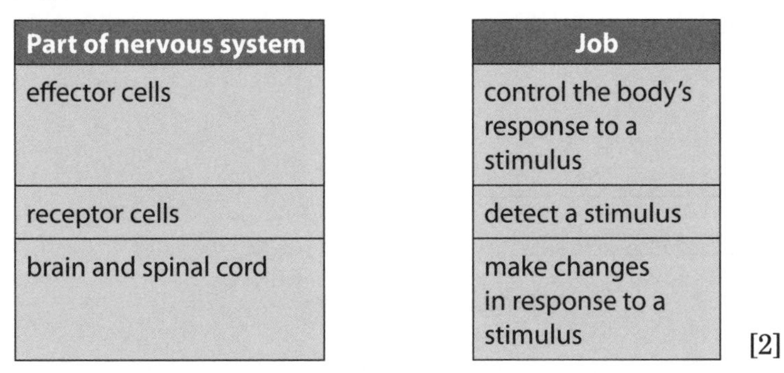

Part of nervous system
effector cells
receptor cells
brain and spinal cord

Job
control the body's response to a stimulus
detect a stimulus
make changes in response to a stimulus

[2]

b Josh is crossing the road. He sees a car coming towards him.

A signal travels from his eyes to his adrenal glands.

His adrenal glands release a hormone, adrenaline.

The adrenaline helps Josh to get out of the way of the car before it hits him.

i The path taken by the nerve signal is shown in the diagram.

Use these words to finish labelling the diagram.

sensory neuron
motor neuron
central nervous system

[3]

2

eye

1

adrenal gland

3

[Not to scale]

ii Finish the sentences. Choose the best words from this list.

Use each word once, more than once, or not at all.

peripheral	**central**
electrical	**chemical**

The signal is carried along

nerve cells by ________________ impulses. The sensory

and motor neurons are part of the ________________

nervous system. The brain and spinal cord form the

________________ nervous system. [3]

Total [8]

5 Dolphins are mammals. They can learn to jump through hoops. Use ideas about repetition, nerve impulses, and neuron pathways to explain how a dolphin's brain changes as it learns to jump through hoops.

____

____

______________________________ [3]

Total [3]

Going for the highest grades

H 6 Synapses are gaps between neighbouring neurons.

a Describe how nerve impulses cross a synapse.

____

____

______________________________ [4]

b Cocaine is an illegal drug. It increases the concentration of serotonin in synapses. It works in a similar way to Ecstasy.

i Suggest how cocaine could cause this change of serotonin concentration.

____

____

______________________________ [2]

ii Suggest why cocaine causes an increase in the transmission of nerve impulses across synapses.

____

____

______________________________ [2]

Total [8]

B7.1 Peak performance – movement and exercise

1 Write the letter **T** next to the statements that are true.
Write the letter **F** next to the statements that are false.

a Bone is living tissue. ________

b Tendons hold the bones together. ________

c Ligaments are inelastic. ________

d Muscles contract to move bones. ________

e Antagonistic muscles always contract together. ________

f The skeleton is only for support. ________

2 Complete the following sentences by crossing out the incorrect word.

Joints contain a lubricant called **sympathetic / synovial** fluid.

A sprain is an overstretched **ligament / tendon**.

Tendons are **elastic / inelastic**.

Cartilage is very **smooth / rubbery**.

A **psychologist / physiotherapist** specialises in joint injuries.

3 Complete the table by writing the function of the component of a joint in each box.

Component of a joint	Function
cartilage	
tendon	
ligament	

4 After an injury Anika was told that the treatment for her injury was RICE. Write down what each letter stands for.

R ________

I ________

C ________

E ________

5 Lewis is starting a regime to improve his fitness. Suggest three questions that his fitness instructor should ask him before he starts exercising.

1 ________________________________

2 ________________________________

3 ________________________________

6 Use words from the box to complete the sentences.

blood pressure	**glucose**	**oxygen**	
decrease	**increase**	**height**	**BMI**

During exercise your heart rate will ____________ .

This is to help deliver more ____________ and ____________ to your muscles.

During an exercise regime, improvement in your physical fitness will make your blood pressure ____________ .

Body mass index (BMI) compares your body mass with your ____________ .

A typical measurement for ____________________ is 120/80 mmHg.

7 Put a tick in the box next to the sentences that are correct.

Physiotherapy will help an injured joint to heal, but it may take weeks. ☐

You should 'run off' an injury such as a sprain. ☐

A fit person will have a lower heart rate than an unfit person. ☐

When measuring your pulse, you should not feel it with your thumb. ☐

The 'pinch an inch' method of measuring body fat is accurate. ☐

The elbow and the shoulder are antagonistic joints. ☐

B7.1.1–2 The skeleton

Humans and other vertebrates have an internal **skeleton**. The skeleton is made from **bone** and **cartilage**.

The skeleton supports the weight of the body, and also allows it to move. As well as its major functions of support and movement, the skeleton also makes blood cells and protects some organs such as the brain.

Joints between the bones allow movement. **Muscles** attached to the bones contract to move the bones.

Muscles act in pairs called **antagonistic pairs** to move bones. One muscle bends the joint (flexing), and the other muscle in the pair straightens the joint (extending).

Remember that muscles can only contract – that is, they get shorter to pull on the bone. They are stretched out to full length again by their antagonistic muscle.

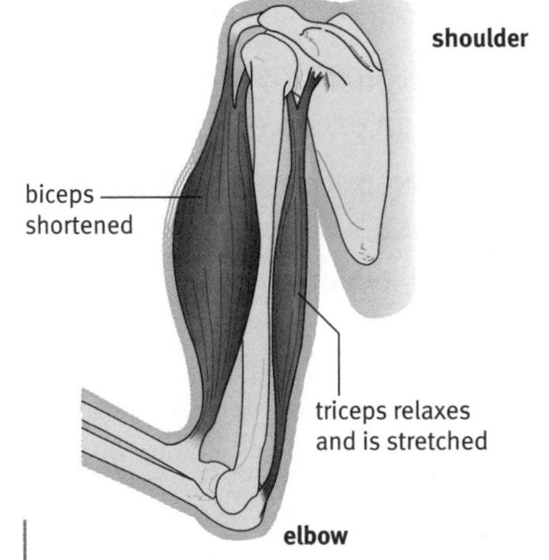

The biceps and triceps muscles contract to move the elbow joint.

B7.1.3–4 Joints

Joints are complex but each component has a specific function to ensure the joint works well. Joints contain **synovial fluid** to act as a lubricant to reduce wear. Synovial fluid works just like oil in a car engine – it makes the movement smoother and reduces wear.

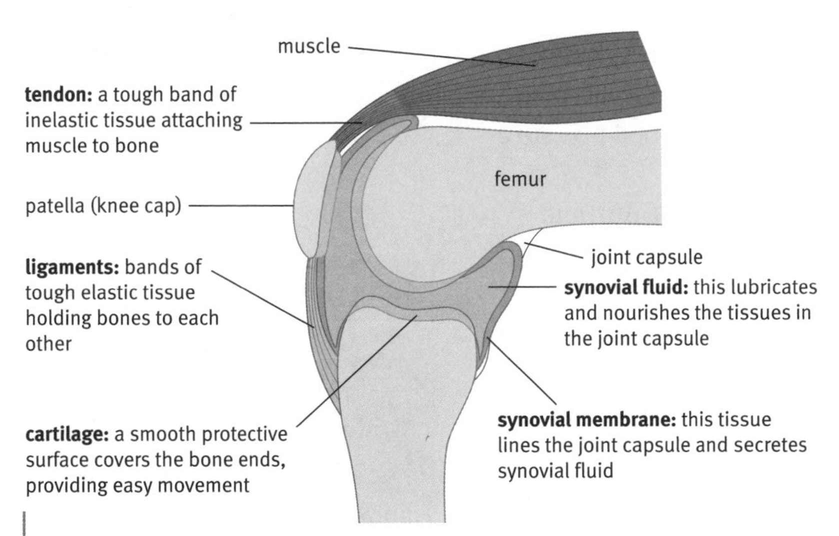

The knee joint is a synovial joint.

The table shows how the components of a joint are suited to their function.

Component of joint	Function in joint	Property
ligaments	hold bones together but allow movement	tough but elastic to allow movement
cartilage	provide protection to ends of bones	very smooth to allow easy movement
tendons	attach muscle to bone	strong and inelastic

Tip

How easily would the joints move if ligaments were not elastic? Think about how your movement might be affected if tendons could stretch!

B7.1.5 Starting an exercise regime

Exercise puts stress on many parts of the body.

Your **medical history** and **lifestyle history** will help a trainer to plan an exercise programme most suitable for you.

Question	Why the information is required
family medical history	Certain circulatory and respiratory conditions run in the family. The trainer can adjust the regime to account for such conditions.
personal medical history	Previous injuries or treatments to muscles and joints can affect your ability to do certain exercises.
medication	You may need medication close at hand (e.g. an inhaler for asthma).
alcohol and tobacco consumption	Both of these can affect performance and progress towards improved fitness.
level of activity	If you are already very active the trainer can start you off at a higher level of exercise.

B7.1.6–8 Showing improvement

Your level of fitness can be assessed in a number of ways:

- **Heart rate** – this increases during exercise to deliver more glucose and oxygen to the muscles. A healthy heart will increase its heart rate less during exercise than an unhealthy heart.
- **Blood pressure** – blood pressure increases during exercise as the heart pumps more forcefully. 120/80 mmHg is a typical value for blood pressure. As you get fitter your blood pressure may not rise as much during exercise.
- **Recovery period** – this is how quickly your breathing and heart rate return to normal after exercise; a shorter recovery time is better.
- **Proportion of body fat** – too much body fat strains the heart and may reduce blood flow in the arteries.
- **Body mass index** (BMI) – this compares your body mass with your height. The BMI indicates whether you are underweight (value below 18.5), healthy (value between 18.5 and 24.9), or overweight (value above 25).

The trainer may measure these at the start of an exercise programme to use as **baseline** data. This is a starting point and your improvement during the exercise regime can be monitored against the baseline data by repeating these measurements at intervals.

Heart rate

Your resting heart rate may be about 70 bpm. As you get fitter this will decrease as the heart becomes more efficient. A highly trained athlete may have a resting heart rate of around 50 bpm.

Calculating BMI

$$\text{BMI} = \frac{\text{body mass (kg)}}{\text{height squared (m}^2\text{)}}$$

Worked example:

For a boy who is 1.75 m tall and weighs 60 kg:

$$\text{BMI} = \frac{60\text{ kg}}{(1.75\text{ m})^2} = 19.6\text{ kg/m}^2$$

B7.1.9 Gathering suitable data

For any data to be meaningful it must be **accurate** and **repeatable**.

Accurate data are close to the true or 'real' value.

Equipment used to measure heart rate, blood pressure, body mass, and so on must be fault free and accurate. Equipment used by medical and sports professionals is checked or calibrated regularly. Always ensure that any equipment is used correctly, does not show signs of wear, and has sufficient power (if needed).

Repeatable means that several runs of an experiment will each produce similar results. Always repeat measurements at least twice.

B7.1.10–11 Injury

Injuries such as **sprains**, **dislocations**, **torn ligaments**, and **torn tendons** can be caused by excessive exercise.

Training tip

Warming up before exercise and gradually building up to higher levels of intensity will reduce the risk of injury.

Sprains

A sprain is caused by overstretching a ligament. The symptoms include: redness and swelling, surface bruising, difficulty in walking, and dull throbbing ache or sharp pains.

The usual treatment is **RICE**:

- **R**est – immobilise the injured joint.
- **I**ce – holding ice over the joint reduces swelling and pain.
- **C**ompression – a bandage around the joint can reduce swelling.
- **E**levation – raising the injured joint helps drain excess fluid away.

B7.1.12 Physiotherapy

Physiotherapists are highly trained and well-qualified health professionals. They specialise in joint and muscle injuries.

After an injury a physiotherapist can help to explain what has happened. They will treat the injury using specific exercises and manipulation to ensure the tissues heal properly and do not shrink or tighten too much. Further exercises designed to strengthen the healing tissue will be suggested for you to do at home. These will help your joint to recover fully, regain its strength, and be less likely to experience the same damage again.

Physiotherapy involves many repeated exercises and can be tedious – some injuries may need several weeks of treatment. It's important to continue the treatments and exercises as suggested or the joint may become permanently stiff or painful.

Use extra paper to answer these questions if you need to.

1 State three functions of the skeleton.

2 How do muscles cause movement?

3 a What is meant by antagonistic muscles?
b Explain why muscles work in antagonistic pairs.
c Name a pair of antagonistic muscles in the arm.
d Which muscle bends the arm?
e Name the muscle that straightens the arm.

4 a Why is it important that bone is living tissue?
b Suggest how exercise might alter the skeleton.

5 a Why do skeletons have joints?
b Where in your body will you find a hinge joint?
c What type of joint is the shoulder?

6 a List six components of a joint.
b Why is cartilage very smooth?
c Describe the function of the cartilage.
d What is the name of the fluid in a joint?
e Describe the role of that fluid.

7 a What is the function of a tendon?
b Explain why it is important that tendons are inelastic.

8 a What is the function of a ligament?
b Explain why ligaments must be elastic.

9 A trainer will ask questions before starting to plan an exercise regime.
a Why will the trainer ask about family medical history?
b How might being a smoker affect an exercise regime?
c Why should a trainer know about previous injuries or treatments?
d Why might the trainer ask about medications that you may be taking?

10 a What is meant by baseline data?
b How will a trainer use your baseline data?
c List three ways that a trainer might measure your current fitness.
d What is meant by the recovery period?

11 a Where is the best place to measure your pulse?
b Why should you use a finger rather than your thumb to measure your pulse?
c At what proportion of your maximum heart rate should you train?

12 a What apparatus is used to measure blood pressure?
b What is a typical healthy value for blood pressure?

13 a Write the equation used to calculate BMI.
b How would you measure your BMI?
c What is considered a healthy BMI?
d How would you measure your proportion of body fat?

14 a List four common injuries that can result from exercise.
b What is meant by a sprain?
c Suggest why football is more hazardous than many other sports.
d Describe the symptoms of a sprain.
e What is the first step in treating a sprain?
f How does applying ice to an injured joint help?
g Explain why a bandage placed around the injured joint should be neither too tight nor too loose.
h Elevation is often recommended after an injury – explain how it helps recovery.

15 a What level of qualification is needed to train as a physiotherapist?
b Why is it important for a physiotherapist to assess an injury before starting treatment?
c A physiotherapist may manipulate a joint. What is meant by manipulation?
d Explain why a physiotherapist will suggest a range of exercises to do at home. What is the aim of these exercises?
e Suggest why a physiotherapist will carefully explain the reasons for doing certain exercises.

H **16 a** What is meant by accuracy?
b Explain why equipment used to measure features such as body fat and blood pressure must be accurate.

17 a What is meant by reliability?
b How can you make your measurements of BMI, heart rate, and recovery period more reliable?

18 How do you calculate a mean?

1 Label the diagram of a knee joint.

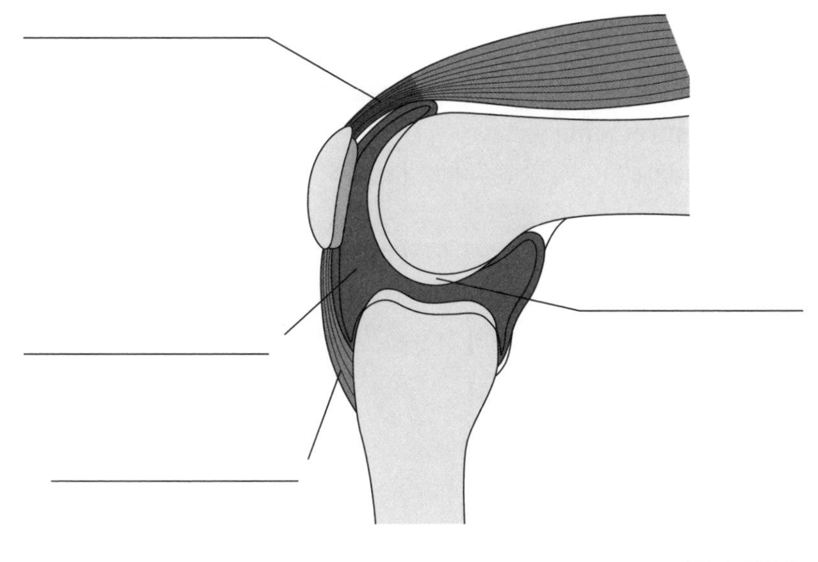

Total [4]

2 David, Peter, and Josh decided to get fit. Before they started they took the following measurements:

- height
- weight
- resting heart rate.

Their results are shown in the table.

	David	Peter	Josh
height (cm)	176	190	183
weight (kg)	82	76	78
resting heart rate (bpm)	72	67	71

a Using the formula below calculate Peter's body mass index (BMI).

Show your working.

$$\text{BMI} = \frac{\text{body mass (kg)}}{\text{height squared (m}^2\text{)}}$$

Answer: ________________ [2]

b Look at the table opposite.

In what category is Peter's BMI?

_______________ [1]

BMI	Category
less than 19	underweight
19–24	normal
25–29	overweight
30–40	obese
over 40	very obese

c David and Peter decided to use their measurements as baseline data. What is meant by baseline data?

____

____

_______________ [2]

Total [5]

H **3** Josh suggested that they should repeat their measurements of resting heart rate each week to monitor their progress. He suggested that they should use a pulse meter and measure their pulse three times.

a Give one reason why using a pulse meter would be better than counting their pulse rate manually.

_______________ [1]

b Give two reasons why they should measure their pulse rate three times.

____

_______________ [2]

On the second day of their training David hurt his ankle, which became swollen and painful.

c Suggest what the injury might be and how it was caused.

____

_______________ [2]

Total [5]

Going for the highest grades

4 Describe how David's ankle should be treated immediately and what role a physiotherapist might have in his recovery.

✎ The quality of written communication will be assessed in your answer to this question.

Write your answer on separate paper or in your exercise book.

Total [6]

B7.2 Peak performance – circulation workout

1 Use the words in the box to complete the sentences below.

carbon dioxide	**urea**	**oxygen**	
two	**oxygenated**	**body**	**glucose**

A double circulatory system has ______________ separate circuits.

One circuit leads to the lungs where the blood is

________________ .

The other circuit carries blood to the ______________ .

Blood carries ______________ in the red blood cells. It also carries ____________ . Both of these substances are needed by the muscle cells.

Blood also carries wastes such as _________ __________ and

__________ .

2 Draw lines to match each component of blood to its function.

Component
red blood cells
white blood cells
platelets
plasma

Function
fight infections
transports dissolved glucose
carry oxygen
help clot blood

3 Explain why red blood cells have each of the properties shown below.

a Red blood cells are filled with haemoglobin.

Explanation: ____

____

b Red blood cells have no nucleus.

Explanation: ____

____

H **c** Red blood cells have a biconcave shape.

Explanation: ____

____

4 Complete these sentences by crossing out the incorrect words.

Arteries / veins carry blood at high pressure.

Blood flows from the **vena cava / pulmonary vein** into the left atrium.

Blood flows from the right ventricle into the **aorta / pulmonary artery**.

The **left / right** ventricle pumps blood to the body.

The aorta carries **oxygenated / deoxygenated** blood.

The **arteries / veins** contain valves.

5 List three properties of capillaries:

a ______________________________

b ______________________________

c ______________________________

H **6** Tick the box next to each substance that passes through capillary walls into the tissue fluid.

plasma ☐

glucose ☐

proteins ☐

carbon dioxide ☐

red blood cells ☐

oxygen ☐

amino acids ☐

7 Put a (ring) around **T** or **F** to show whether each sentence is **true** or **false**.

A	All mammals have a double circulation.	T	F
B	All arteries carry oxygenated blood.	T	F
C	The brain has its own separate circuit in the double circulation.	T	F
D	Blood carries oxygen and carbon dioxide.	T	F
E	The right ventricle is thicker than the left ventricle.	T	F
F	Valves make blood flow in the right direction.	T	F

8 Place the statements in the correct order. Write the letters in order in the boxes below. The first one has been done for you.

A Blood is pumped out of the ventricles.

B The atria contract.

C Blood flows into the atria.

D Blood passes from the atria to the ventricles.

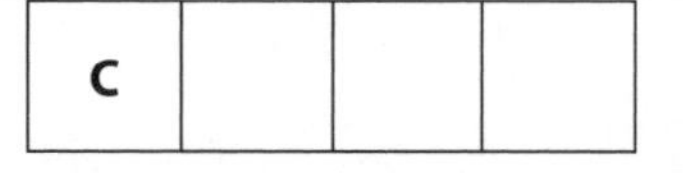

C			

B7.2.1–2 The circulatory system

Mammals have a **double circulatory system**.

- The blood passes through the heart twice for each circulation around the body.
- The heart acts as two separate pumps.
- Blood from the left side of the heart goes around the body and returns to the right side.
- Blood from the right side of the heart goes to the lungs and returns to the left side.

The role of the circulatory system is to transport materials around the body.

Oxygen and glucose are transported to the muscles.

Wastes such as carbon dioxide are transported away from the muscles.

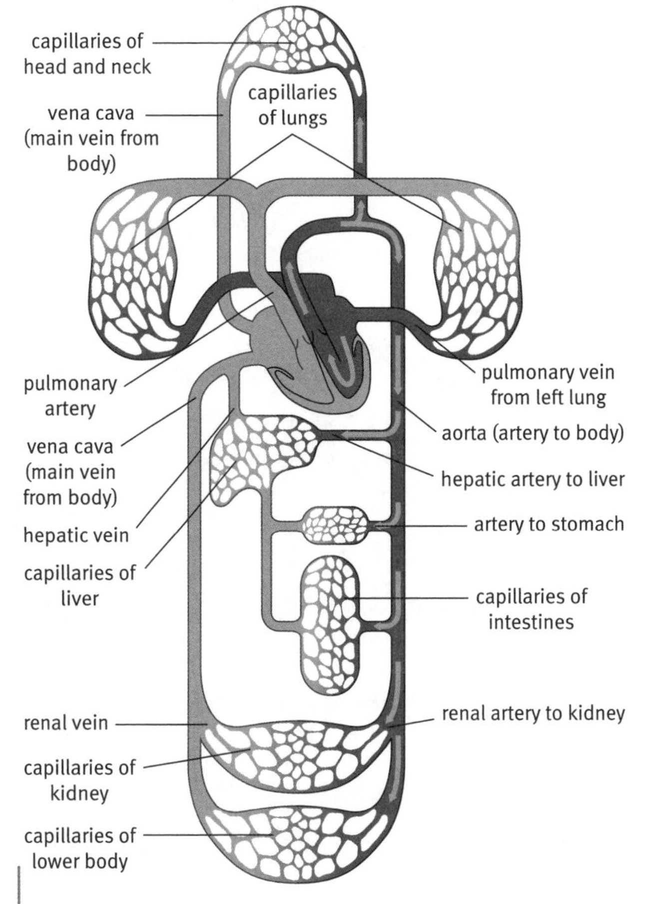

The double circulation.

B7.2.3–4 Blood

Blood is a complex tissue containing different types of cell in a fluid called plasma. Many other substances such as glucose are dissolved in the plasma.

- **Red blood cells** transport oxygen.
- **White blood cells** help to fight infections. Some white blood cells make antibodies, and others engulf and digest microorganisms in a process called **phagocytosis**.
- **Platelets** help to clot blood at sites of injury by sticking to the cut edge and releasing chemicals that cause blood to clot.
- **Plasma** transports nutrients (such as glucose and amino acids), antibodies, hormones, and wastes (such as carbon dioxide and urea).

The red blood cells are very important in sporting performance – they carry the oxygen to the muscles. Red blood cells are especially adapted to ensure they carry a lot of oxygen.

Feature of red blood cell	How it helps to carry oxygen
packed with haemoglobin	oxygen binds to haemoglobin
no nucleus	more space inside cell for more haemoglobin
biconcave shape	increases surface area for rapid exchange of oxygen

B7.2.5 The heart

The heart is a double pump.

The left side pumps **oxygenated** blood to the body delivering oxygen and nutrients to the cells.

Tip

Remember that a double circulation is just that – two separate circuits, one to oxygenate the blood in the lungs, the other to deliver materials around the body.

Tip

Blood is a tissue – a number of types of cell all serving one function.

Some blood statistics

We have about 6 litres of blood in our bodies. There are about 5 million red blood cells per mm^3 of blood – that means we have 30 million million red blood cells!

The right side pumps **deoxygenated** blood to the lungs to remove carbon dioxide and collect oxygen – a process called oxygenation.

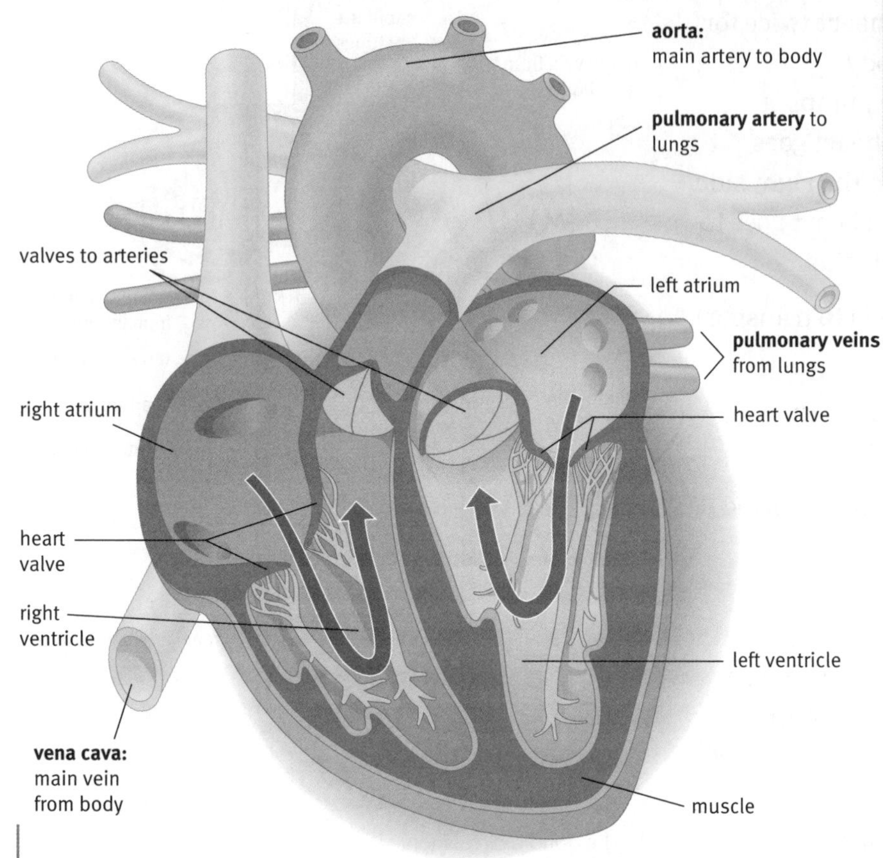

The human heart. 'Left' and 'right' refer to the way your heart lies in your body.

Tip

Do not use the term 'oxidised' for blood as this refers to a chemical reaction. Blood carrying oxygen is oxygenated.

The heart consists of four chambers.

- The upper chambers are called atria (singular is **atrium**) and they receive blood from the veins.
- The lower chambers are called **ventricles** and they pump blood out into the arteries. Therefore ventricles have thicker muscles than atria.
- Blood in the left ventricle is pumped out through the **aorta** and around the body. So the left ventricle has the thickest muscle. It returns through the **vena cava** to the right atrium, which pushes it into the right ventricle. From here the blood is pumped through the **pulmonary artery** to the lungs before returning through the **pulmonary vein** to the left atrium.

You should remember that while the left and right sides of the heart are separate, both atria pump at the same time and both ventricles also pump together.

Tip

Remember that veins always bring blood back to the heart and arteries always carry blood away from the heart.

Tip

The arteries carry blood at high pressure and do not need valves. Veins carry blood at low pressure and here valves are needed to stop blood running down and pooling in the feet and hands.

B7.2.6 Valves

Valves prevent blood from flowing backwards.

The heart has two sets of valves:

- between the atria and the ventricles; these stop blood flowing from the ventricles back to the atria

- between the ventricles and the arteries; these stop blood flowing from the arteries back to the ventricles

There are also valves at intervals along the veins. These valves stop blood flowing backwards.

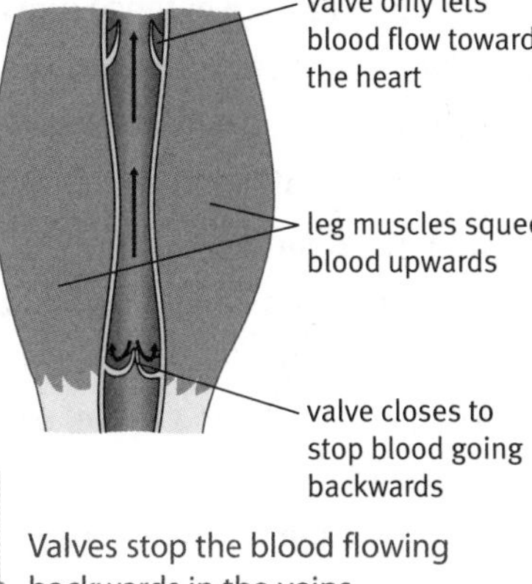

Valves stop the blood flowing backwards in the veins.

B7.2.7 Capillaries and exchange

All organs contain a network of **capillaries** in a **capillary bed**.

- These tiny vessels create a large surface area to help exchange of materials into and out of the blood.
- They have a very thin wall (just one cell thick) to enable easy exchange.
- Capillary walls are porous – they allow some of the plasma to leak out into the surrounding tissue. This fluid becomes **tissue fluid**. It contains all the chemicals that were dissolved in the blood plasma. So the tissue fluid contains all the oxygen and glucose needed by the muscle cells.
- The substances needed by cells diffuse from the tissue fluid into the cells. Waste substances such as carbon dioxide diffuse out of the cells into the tissue fluid.
- The fluid then returns to the capillary carrying the waste substances.

Tip

Do not say that capillary walls are leaky – use the term porous.

The role of blood pressure

As blood flows from the artery into the capillary bed it is still at high pressure. This pressure pushes the fluid out through the porous capillary walls.

The capillary is so tiny that that there is a lot of resistance to flow and the pressure drops along the capillary.

At the vein end of the capillary the pressure is low, allowing the tissue fluid to return to the capillary.

Tissue fluid helps nutrients and waste products to be exchanged by diffusion in the capillary beds.

B7.2: Peak performance – circulation

Use extra paper to answer these questions if you need to.

1 a Why do we need a circulatory system?
b Name the three types of blood vessel.
c Which type of blood vessel carries blood away from the heart?
d Name two substances carried by the blood to the muscles.
e Which type of blood vessel carries blood back to the heart?
f Name two things carried away from active muscles.

2 a Describe the route taken by blood from the lungs to the body and back to the lungs.
b What is the name given to this type of circulation?
c Where is the highest blood pressure found?
d Where is the lowest blood pressure found?

3 a Suggest why mammals need a double circulatory system.
b Suggest what advantages it provides over a single circulatory system.

4 a List the components of blood.
b What makes red blood cells red?
c Name five substances transported in blood.
d Name the gas transported to muscles in red blood cells.

5 a How much blood is found in a typical adult human body?
b How many red blood cells are there in blood, per mm^3?
c Use these figures to calculate how many red blood cells you have in your body.

6 a State the role of white blood cells.
b What happens during phagocytosis?
c What is the function of the platelets?

7 a Name the substance in red blood cells that binds oxygen.
b Explain why red blood cells have no nucleus.

H **8 a** State the name used to describe the shape of a red blood cell.
b Explain why red blood cells are this shape.

9 a What is the role of the heart?
b What is a typical heart rate at rest?
c What happens to the heart rate when we exercise?

10 a Name the upper chambers in the heart.
b Name the lower chambers in the heart.
c Which side of the heart pumps blood to the lungs?
d Which chamber has the thickest wall?
e Explain why it needs the thickest wall.
f Why do the atria have very thin walls?

11 Name the vessel that:
a carries blood into the left atrium
b carries blood out to the body
c carries blood out to the lungs
d carries blood into the right atrium.

12 a Which contracts first – the atria or the ventricles?
b Describe the cycle of events in one heartbeat.
c What makes the lub-dub sound of the heart?

13 a Why does the heart have valves?
b What is the role of the valves between the atria and the ventricles?
c What is the role of the valves between the ventricles and the arteries?
d Which arteries (if any) have valves?
e Explain why veins need valves.
f Describe how pressure is created in the veins to move the blood upwards against gravity.

H **14 a** Where in the body are capillaries found?
b What is the role of the capillaries?
c Why are there such large numbers of capillaries?
d Name the vessel that brings blood to a capillary bed.
e Name the vessel found at the other end of the capillary bed.
f Which end of the capillary carries blood at the higher pressure?
g Explain why blood pressure drops in the capillaries.

15 a List three properties of capillaries that help make exchange efficient.
b Name the components of blood that do not leave the capillaries.
c Name three substances in plasma that pass out of the capillary.
d At which end of the capillary do substances leave the blood?
e What is the name for the fluid formed outside the capillary?
f Name two substances that pass back into the capillary.
g At which end of the capillary do substances re-enter the blood?
h Explain the role of tissue fluid in supplying cells.

1

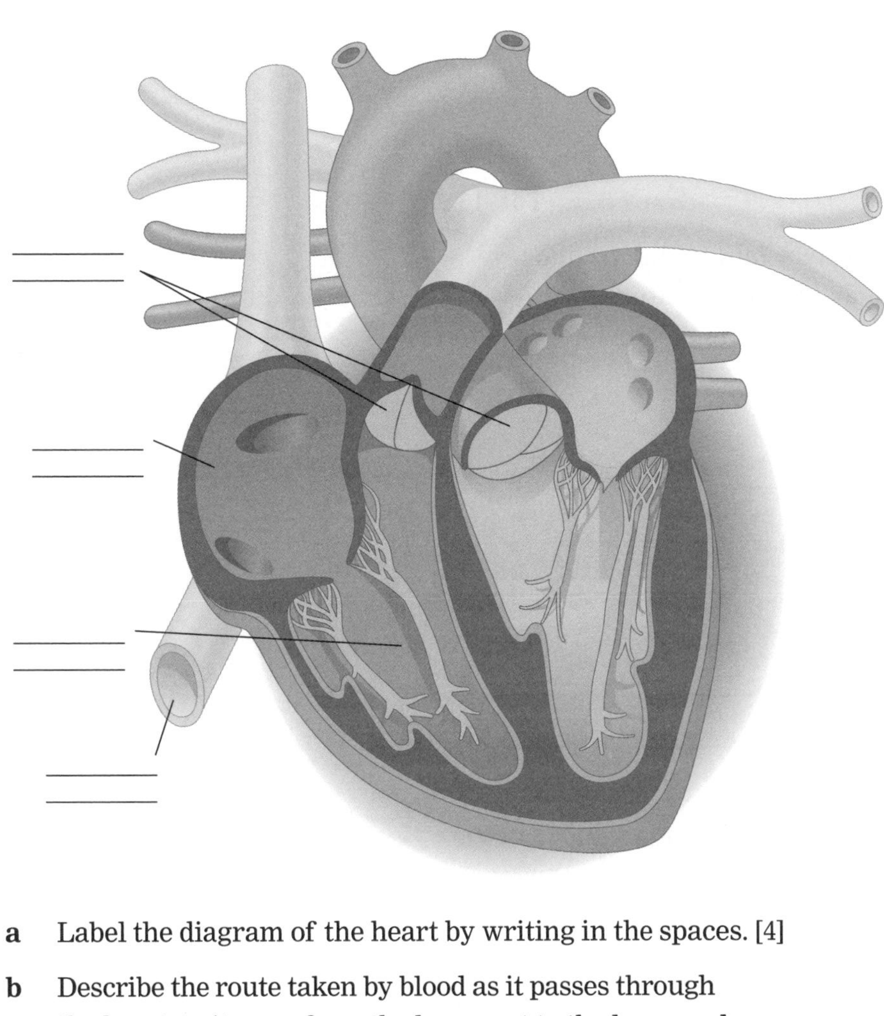

a Label the diagram of the heart by writing in the spaces. [4]

b Describe the route taken by blood as it passes through the heart on its way from the lungs out to the leg muscles.

____

____

______________________________ [3]

c Oxygenation happens in the lungs. What is meant by oxygenation?

____

______________________________ [2]

d Name the only artery that carries deoxygenated blood.

______________________________ [1]

Total [10]

2 a Complete the table below.

Component of blood	Function
red blood cells	
white blood cells	
platelets	

[3]

H

b Explain why red blood cells have a biconcave shape.

____

____ [2]

Total [5]

3 Blood pressure does not remain constant in the circulatory system. Use your knowledge about blood pressure in different parts of the system to explain the following features of the heart and circulatory system.

a The atria have thin walls.

____

____

____ [2]

b The veins contain valves.

____

____

____ [2]

c The left ventricle has a thicker wall than the right ventricle.

____

____

____ [2]

Total [6]

H

4 Capillary beds enable exchange to occur between the blood and the body cells.

a What features of a capillary bed help make exchange efficient?

____

____

____ [2]

b State three differences between tissue fluid and blood.

i ____

ii ____

iii ____ [3]

Total [5]

Going for the highest grades

5 Describe how tissue fluid is formed and how this process contributes to the exchange of materials at the capillaries.

The quality of written communication will be assessed in your answer to this question.

Write your answer on separate paper or in your exercise book.

Total [6]

1 Below are some statements about body temperature. Tick the boxes next to the statements that are correct.

Your temperature may be lower at night. ☐

Your extremities may be cooler than your core. ☐

You gain heat if your environment is warmer than you. ☐

Putting more clothes on will increase insulation. ☐

Sweat cools you down by evaporation. ☐

Shivering makes you lose more heat. ☐

H Vasoconstriction makes your skin look red. ☐

2 Complete the following sentences.

Your body temperature should be around ________________ °C.

You can generate heat in your body by using sugar in the process of ________________ .

If you are too cold the blood vessels to the skin will __________ .

Behaviour such as fanning yourself helps you to

________________ ________________ .

H Body temperature is detected in the ________________ of the brain.

3 Put a ring around **T** or **F** to show whether each statement is **true** or **false**.

A	The skin contains temperature receptors.	**T**	**F**
B	The muscles contain temperature receptors.	**T**	**F**
C	The muscles can generate heat.	**T**	**F**
D	Hairs on the skin become erect when you are cold.	**T**	**F**
E	Blood transfers heat around the body.	**T**	**F**
F	Less blood goes to the skin when you are hot.	**T**	**F**
G	H Vasodilation and vasoconstriction are antagonistic actions.	**T**	**F**
H	The hypothalamus is a processing centre.	**T**	**F**

4 Complete the table to compare type 1 and type 2 diabetes.

Type 1 diabetes	Type 2 diabetes
Often appears in young people.	
	Cells in the body do not respond to insulin.
	Treated by careful diet and exercise.

5 Put a (ring) around **G** or **B** to show whether each lifestyle choice is **good** or **bad**.

A	eating fast food such as burgers every day	**G**	**B**
B	walking to school every day	**G**	**B**
C	eating a large bar of chocolate on the way home	**G**	**B**
D	making every journey in a car	**G**	**B**
E	eating wholegrain rather than refined rice and pasta	**G**	**B**
F	eating an orange rather than sweets	**G**	**B**
G	spending all your free time playing games on a computer	**G**	**B**
H	joining a sports club	**G**	**B**

6 Explain each of the following observations.

a When John started regular exercise classes he lost weight.

____

____

____

b Avril always seems to feel very tired about an hour after a sugary meal.

____

____

____

c People who eat lots of fruit and vegetables have a lower risk of stomach cancer.

____

____

____

B7.3.1 A balancing act

We need to maintain our body temperature at about 37 °C.

Heat can be gained in the body in two ways:

- from respiration in the cells
- from the environment, if the air is warmer than 37 °C.

Heat can be lost from our skin to the environment on cool days.

Our **core** body temperature is most important. **Extremities** such as hands and feet may be cooler.

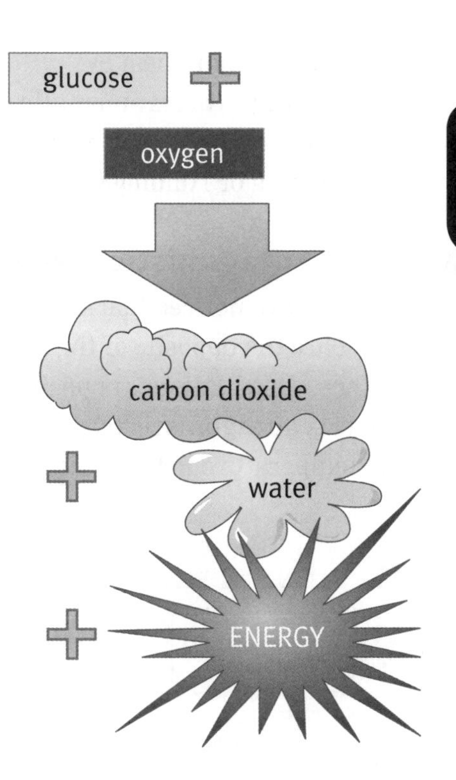

B7.3.2–5 Monitoring temperature

The skin contains temperature receptors that detect external temperatures.

An area of the brain contains temperature receptors that detect the temperature of the blood. It also acts as a **processing centre** – it receives information from the temperature receptors and sends instructions to the effectors. The effectors include the sweat glands and the muscles.

This area of the brain that detects and processes information about temperature is the **hypothalamus**.

B7.3.6–7 Keeping cool

On a hot day and when you are active your body may gain heat and start to warm up. If this happens your body will respond to increase heat loss so that core body temperature is not affected.

- Heat is produced during respiration in cells – so when you exercise you start to feel hot. Reducing your activity level will reduce heat production.
- Sweat is made in **sweat glands** in the skin. It comes to the surface of the skin and **evaporates**. This evaporation helps to remove excess heat.
- When you get hot you may also look red. This is due to a process called **vasodilation**. When you need to lose more heat your blood is diverted to run close to the surface of the skin so that it can lose more heat.

Tip

Remember that it is the evaporation of sweat that cools you down. If the sweat does not evaporate you will not get cooler.

Vasodilation

The blood vessels near the surface of the skin are filled with blood. Energy from the warm blood is transferred down the temperature gradient to the environment.

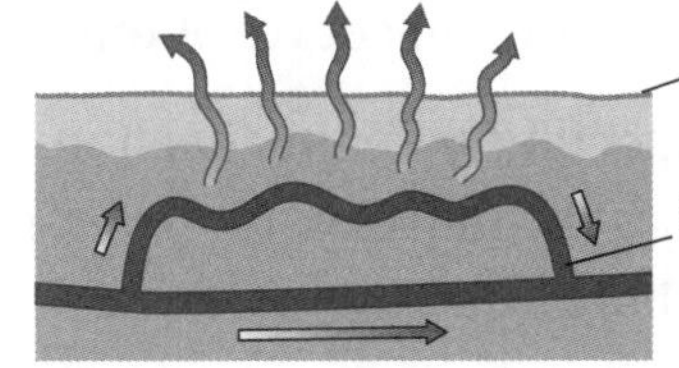

surface of skin

blood vessels near surface of skin

Key

flow of blood in blood vessels

energy loss from skin surface

B7.3.8 Keeping warm

On a cold day or when you are inactive you may not gain enough heat to match the losses and your body starts to cool down. Your body will respond to reduce heat loss and create more heat so that core body temperature is not affected. You can conserve heat:

- by putting on more clothes or curling up
- by the hairs on your skin standing erect to trap more air as insulation.
- When you get cold the blood is diverted away from the surface of the skin by constriction of the blood vessels leading to the surface. This is called **vasoconstriction**. It means that the blood does not lose as much heat and your skin may look paler.

Vasoconstriction

The muscles in the walls of blood vessels near the surface of the skin contract. Less blood flows near the surface of the skin, so less energy is lost to the environment.

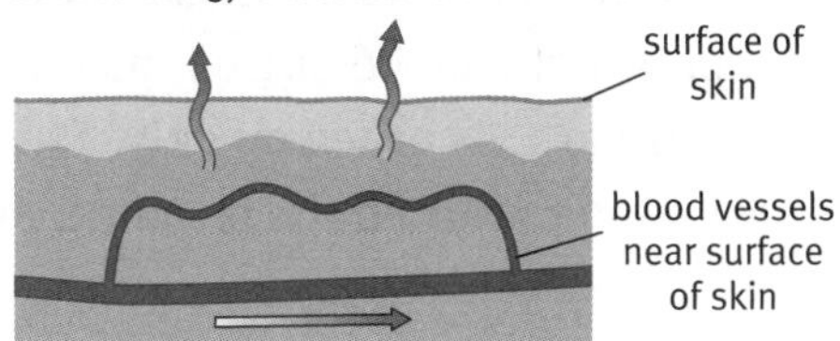

surface of skin

blood vessels near surface of skin

You can make more heat by:

- **shivering** – this is automatic contraction and relaxation of your muscles to speed up respiration
- exercising or rubbing your hands and arms.

B7.3.9 Antagonistic responses

You may have noticed that some of the responses to be being too hot are the opposite of the responses to being too cold. For instance, vasodilation is opposite to vasoconstriction. These are known as **antagonistic** responses. This allows the control of body temperature to be very sensitive and precise.

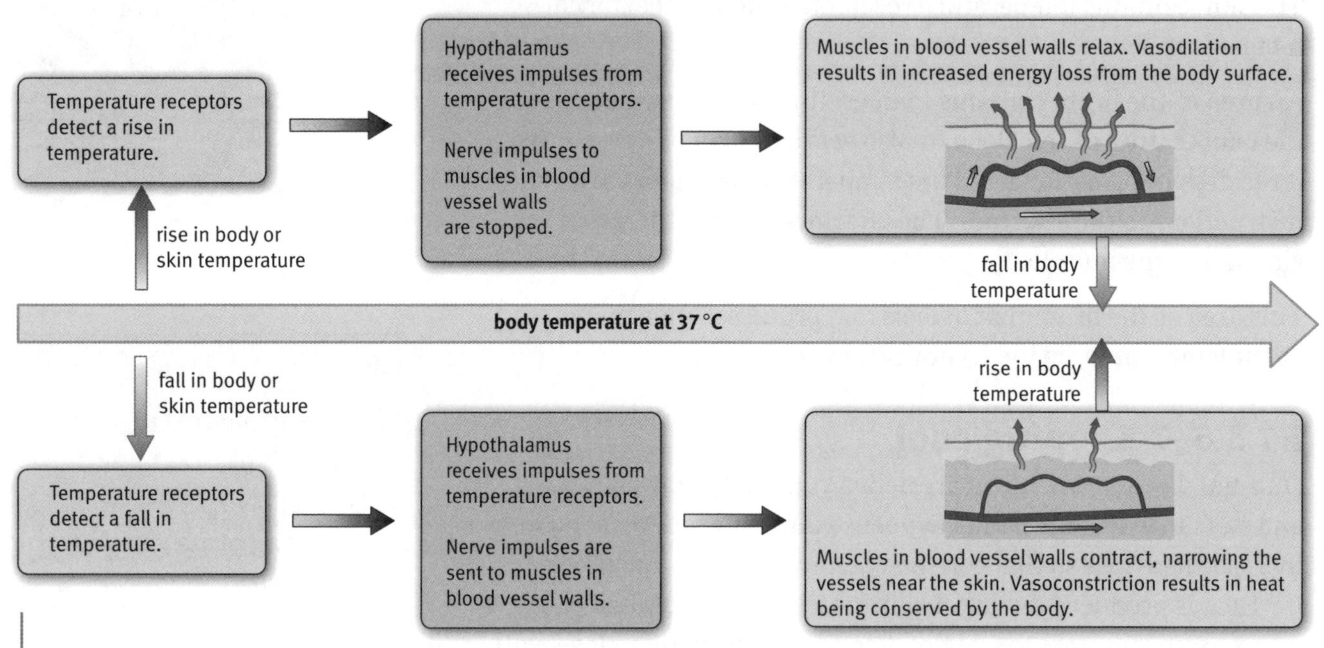

Temperature control in the body is brought about by antagonistic responses.

B7.3.10 Sugar balance

Sugar is the main source of energy for respiration.

It is important to maintain enough sugar in the blood to provide energy to the muscles. However, too much sugar in the blood can be harmful. Therefore the body must balance its blood-sugar concentration.

Blood-sugar levels are controlled by the hormone insulin, released by the pancreas. If blood sugar is too high then insulin is released to reduce the blood-sugar level.

Diet is an important aspect of sugar balance. High levels of sugar intake, such as those found in many processed foods, will increase the blood-sugar concentration quickly because simple sugars are absorbed quickly. This means extra insulin has to be released to return the blood sugar to normal.

Tip

Remember that insulin is a hormone. It causes the liver to take sugar out of the blood.

B7.3.11–13 Diabetes

People with diabetes cannot control their blood-sugar level.

There are two types of diabetes.

Diabetes	Cause	Symptom	Treatment
Type 1	The pancreas does not make enough insulin. Can occur in young people.	Dangerously high blood sugar after a sugary meal. Thirst, large volumes of urine.	Regular injections of insulin. Carefully monitored sugar intake.
Type 2	Insulin is produced but the cells cannot use it properly. Poor diet, inactive lifestyle, obesity. Usually occurs in middle age, but can sometimes be seen in younger people.	Thirst, frequent urination, and tiredness. Over longer periods the arteries, kidneys, and eyesight may be damaged.	Control of diet to reduce sugar intake. Regular exercise to use up sugars.

B7.3.14–15 Exercise and diet

It is important to understand the effects of exercise and a healthy diet.

- Exercise uses energy in the form of sugars taken from the blood. Therefore regular exercise will help to reduce blood-sugar levels that have become too high.
- It is better to prevent the blood-sugar level getting too high in the first place. A meal high in refined sugars such as many fast-food meals or ready-made meals will quickly raise the blood-sugar level.
- Food containing complex carbohydrates such as pasta, bread, and rice takes longer to digest and releases the sugar more slowly. This allows the body to store the sugar correctly or use it during exercise before the blood-sugar concentration gets too high.
- There are a number of diseases that are linked to poor lifestyle choices such as poor diet and lack of exercise.
- These include heart disease, **obesity**, diabetes, tooth decay, anorexia, and a range of **cancers**, especially bowel cancer.
- Scientists study patterns of these diseases and look for factors that appear to increase the risk of getting the disease – these are called **risk factors**.

Tip

Remember that to maintain your body weight energy intake must equal energy use.

Tip

Remember that a risk factor does not cause the disease on its own – it simply increases the chances that you will develop the disease.

Use extra paper to answer these questions if you need to.

1 a Name the process that releases heat inside the body.
b What is the energy source for this heat?
c Name a gas that is needed to release the heat.
d Name two products of the reaction other than heat energy.
e Why do you feel hot after exercise?
f Name the organs in the body where a lot of heat is produced.

2 a What is the normal human body temperature?
b Explain why the extremities may be cooler.
c Describe how heat is transferred around the body.
d List the places in the body where temperature receptors are found.

3 a What is an effector?
b Name two effectors used in controlling body temperature.
c Why is your body slightly cooler at night than during the day?
d What happens to body heat if your body is warmer than the surroundings?
e Suggest three types of behaviour that will reduce heat loss.

4 a Explain why shivering will help to keep you warm.
b Suggest two automatic responses within the body other than shivering that will help to keep you warm.

5 a How does your body respond if you are too hot?
b Explain how sweating helps you to cool down.
c Explain why sweat does not cool you down if it cannot evaporate.
d Suggest weather conditions that would make it difficult for sweat to evaporate.

6 a What happens to blood-sugar levels soon after a sugary meal?
b Name the hormone that controls blood-sugar levels.
c Describe how this hormone controls blood-sugar levels.
d What symptoms might a person feel when they have a low blood-sugar level?

7 a What sorts of food release sugars slowly?
b Name three examples of this type of food.

8 a What is the cause of type 1 diabetes?
b What happens to a person with type 1 diabetes if their blood sugar drops too low?
c Describe two symptoms of type 1 diabetes.
d How is type 1 diabetes treated?

9 a At what age is type 2 diabetes most likely to develop?
b What types of lifestyle are linked to type 2 diabetes?
c How should type 2 diabetes be treated?

10 a List the components of a healthy diet.
b Suggest five serious illnesses that can result from an unhealthy diet.

11 a List three aspects of lifestyle that lead to better health.
b Explain why exercise helps to keep your weight down.
c How else does exercise improve your body?
d List three types of exercise that may help you to lose weight.

12 a Explain why a person with an active occupation needs to eat more.
b Suggest three reasons why a diet of processed food is unlikely to be healthy.

13 a What is a risk factor?
b Name one risk factor associated with each disease:
i heart disease
ii diabetes
iii tooth decay
iv bowel cancer
v mouth and oesophagus cancer.
c Explain why scientists look for links between lifestyle choices and certain diseases.
d Why might countries such as Japan have lower rates of heart disease than the UK?

H

14 Where in the brain is the processing centre to keep your temperature constant?

15 a What is meant by vasoconstriction?
b Explain how vasoconstriction reduces heat loss.
c How does vasoconstriction change the appearance of someone's skin?
d Explain why you look flushed when you are hot.

16 a What is meant by the word 'antagonistic'?
b Describe two processes that are antagonistic in controlling body temperature.
c Explain the benefits of antagonistic processes.

1 The table shows the amount of energy used during different activities.

Activity	Energy (kJ/h)
sitting quietly	1.7
writing	1.7
standing relaxed	2.1
vacuuming	11.3
walking rapidly	14.2
running	29.3
swimming (4 km/h)	33.0

a Which activity uses most energy?

______________________________ [1]

b Calculate the ratio of energy used by a person running compared with a person walking rapidly. Show your working and express your answer to one decimal place.

Answer ______________________________ [2]

c Suggest why a domestic cleaner tends to stay slimmer than a person who spends most of their time watching television.

____

____

______________________________ [2]

Total [5]

H **2 a** Describe the role of the hypothalamus in temperature control.

____

____

______________________________ [3]

b Describe the changes to the skin that occur in response to being too cold.

____

____

____

______________________________ [3]

c Explain the advantage of having effectors that work antagonistically.

____

____

____

______________________________ [2]

Total [7]

3 a i Name the organ that releases insulin.

______________________________ [1]

ii What triggers the release of insulin?

______________________________ [1]

iii Describe the effect of insulin in the body.

____

______________________________ [2]

b Explain why:

i a person with type 1 diabetes needs to inject insulin.

____

______________________________ [2]

ii a person with type 2 diabetes should exercise regularly.

____

______________________________ [2]

Total [8]

4 Read the extract opposite from a scientific paper.

> People who consume kidney beans or lentils at least three times a week reduce their risk of developing bowel cancer by a third. Eating brown rice once a week cuts the risk by two fifths, while having cooked green vegetables at least once a day reduces it by a quarter.

a What are the benefits of eating lentils regularly?

____

______________________________ [2]

b What component of a healthy diet is found in brown rice that may not be in processed rice?

______________________________ [1]

c Suggest why eating cooked green vegetables regularly reduces the risk of bowel cancer.

____

____

______________________________ [3]

Total [6]

5 Describe the advantages to health of regular exercise and a healthy diet.

✎ The quality of written communication will be assessed in your answer to this question.

Write your answer on separate paper or in your exercise book.

Total [6]

1 Complete the following sentences. Use words from the list to fill in the gaps.

enzymes	substrate	open-loop
sustainable	bacteria	erosion

An open-loop system is not ________________ .

An ecosystem is a sustainable system. In a system like this the waste from one process is used as a ____________________ in another process.

Dead organic matter is broken down by ____________________ which use digestive ______________ .

Over-exploitation of soil removes the mineral nutrients and causes ______________ .

2 Put a (ring) around **T** or **F** to show whether each statement is **true** or **false**.

Energy can be recycled.	T	F
Human systems are linear.	T	F
Ecosystems are perfect closed loops.	T	F
Excess production of sperm is wasted.	T	F
Earthworms help to recycle mineral nutrients.	T	F
Deforestation causes flooding.	T	F
Oil is fossil sunlight energy.	T	F
All plastics can be decomposed.	T	F

H 3 Place the sentences in the correct order to describe eutrophication. Write the letters in the boxes to show the correct order.

A This leads to an increased growth of algae.

B Dead algae decay in the water.

C The decay is caused by bacteria that use oxygen.

D There is an increase in mineral nutrients.

E As a result, all the oxygen is used up.

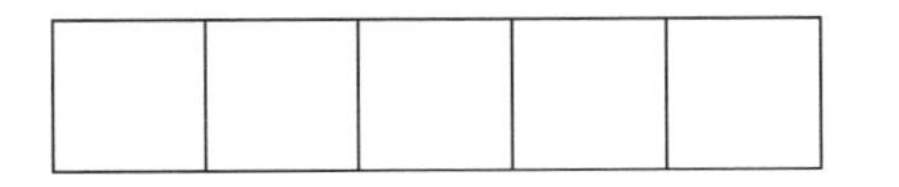

4 Explain the following statements.

a Deforestation can cause desertification.

____

____

b Ecosystems are closed-loop systems.

____

____

c Open-loop systems are not sustainable.

____

____

d Bacteria are essential for plant growth.

____

____

5 Draw lines to match each term to its meaning.

Term
biofuel
fossil sunlight energy
intensive agriculture
open-loop system
quota

Meaning
a limit to the number of fish that can be caught
a system with a lot of waste
energy stored millions of years ago
a human system involving high inputs
fuel made from plant growth

6 **a** Explain what is meant by an 'ecosystem service'.

____

____

b Give three examples of ecosystem services.

____

____

____

7 Explain the following observations.

a Sperm and eggs are often produced in large numbers.

____

____

b Many crops rely on insects.

____

____

c Stable ecosystems are closed-loop systems.

____

____

d An ecosystem can rarely be a completely closed loop.

___

___

e Human activities often destabilise an ecosystem.

___

___

8 Put a tick in the box next to each statement that is correct.

Toxins are more concentrated lower down a food chain. ☐

Energy can be recycled. ☐

Vegetation protects the soil from erosion. ☐

Agriculture reduces biodiversity. ☐

Dead organic matter contains energy. ☐

Mineral nutrients are recycled by bacteria. ☐

All plastics are biodegradable. ☐

B7.4.1–5 Ecosystems as closed-loop systems

An **open-loop system** has an input at one end and an output at the other. It cannot be maintained without continued inputs. It is therefore not **sustainable**.

A **closed-loop system** has no waste. Outputs from one process become inputs for another process.

Ecosystems are closed-loop systems because waste materials are reused within the system. For example:

- Oxygen released from photosynthesis is used in respiration.
- Carbon dioxide released from respiration is used in photosynthesis.
- **Dead organic matter** such as leaves, fruit, and faeces can be eaten by small organisms such as worms or digested by **microorganisms** using **digestive enzymes.**
- Chemicals such as water, nitrogen, carbon, and oxygen move through an ecosystem in cycles. Wastes from one reaction become **reactants** in another.

Reusing waste materials makes an ecosystem sustainable. It can continue to work without inputs of materials. The only input required is energy, which comes from the Sun.

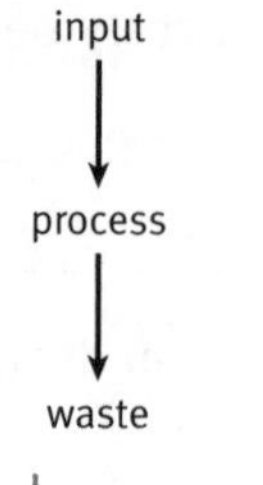

An open-loop system.

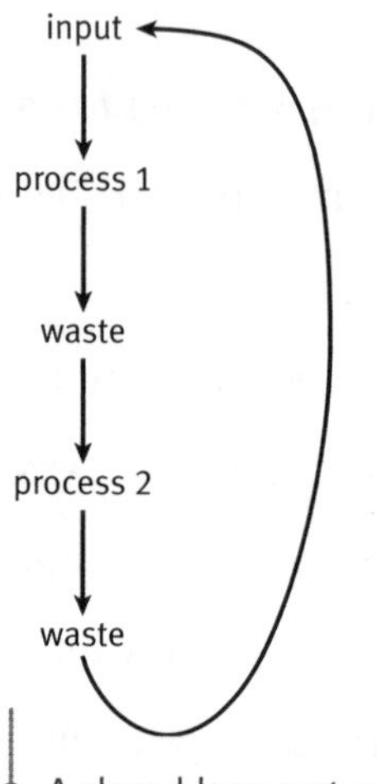

A closed-loop system.

B7.4.6–9 Balance

Not all ecosystems are perfect closed-loop systems. There may be losses such as migration of animals, or nutrients carried away by air currents or water.

In a stable ecosystem, such as a rainforest, these losses are balanced by gains from other ecosystems.

There are examples of apparent waste in natural ecosystems – for example, the apparent overproduction of eggs, sperm, **flowers**, and **pollen**. However, these are essential to ensure that reproduction is successful. Fertilisation and **pollination** occur by chance and producing a lot of eggs and sperm, pollen, flowers, and **fruit** increases the chances of successful reproduction. Continuation of all species through successful reproduction is essential to maintaining a **stable ecosystem**.

Remember that in a stable ecosystem the excess structures produced will be broken down and recycled – so the balance is maintained.

Tip

Remember that the decay of dead organic matter is essential for returning mineral nutrients to the soil.

Tip

Remember that minerals are recycled, but energy cannot be reused – it passes through the ecosystem and is lost.

B7.4.10–11 Vegetation and soil stability

The stability of a natural ecosystem depends on soil stability. Vegetation has an important role in maintaining the soil and preventing **soil erosion**.

- The roots bind the soil together.
- Foliage protects the soil from direct rainfall and from extremes of temperature and can promote cloud formation.

Tip

Erosion is damage to exposed soil. The damaged soil may then be blown or washed away, causing desertification.

B7.4.12–13, B7.4.15 Human influence

Humans make use of natural ecosystems and exploit them for **ecosystem services**. These include the following:

- Providing clean air – plants use carbon dioxide and restore oxygen levels.
- Providing clean water – the natural water cycle involves rainfall that soaks into the soil, providing plants with water that then evaporates from the leaves or slowly drains into rivers. **Deforestation** of large areas breaks the natural cycle and causes flooding or drought. Maintaining forests can restore the natural balance, supplying water to human populations.
- Soil to grow crops – soil is maintained by the organisms that live in the soil. Breaking their natural cycles destroys the soil. Processes such as ploughing can do this.
- Mineral nutrients for crop growth – continuous cropping removes all the minerals and disrupts the natural cycles.
- Pollination – many crops are pollinated by insects.
- Fish and game – we remove animals from the cycle for food. Over-use will disrupt the natural cycle.

Human exploitation is often harmful. Our current 'take–make–dump' system produces a lot of waste such as non-recycled household goods, emissions from burning fossil fuels, and excess materials from intensive agricultural practices. Human systems are often not closed-loop systems. These practices unbalance natural ecosystems by altering the inputs and outputs. This is not sustainable.

Tip

Humans have always used ecosystems. In the past we have lived within the ecosystem. However, now we are over-exploiting and trying to take too much out of the loop.

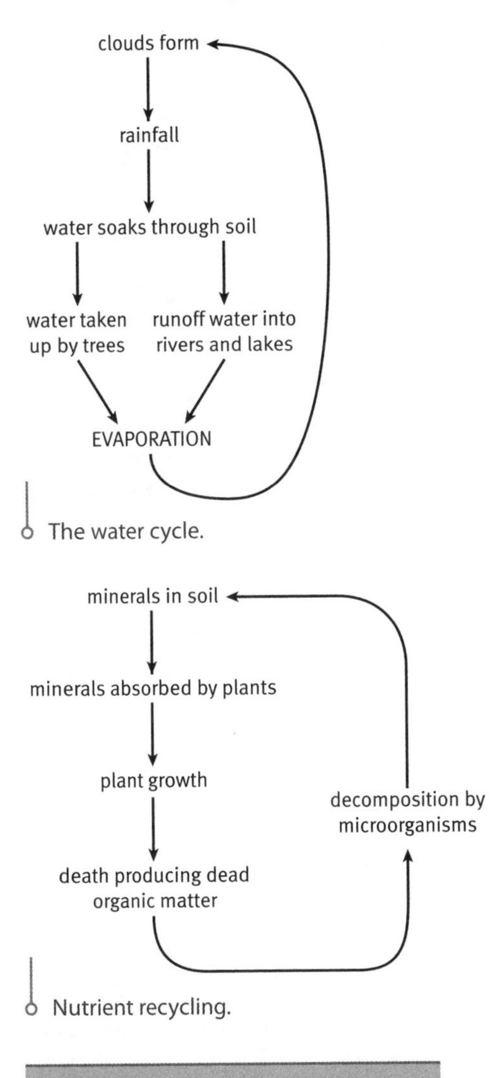

The water cycle.

Nutrient recycling.

B7.4.14, B7.4.16–20 How humans harm natural ecosystems

- Non-recycled waste is dumped or buried in groundfill sites. Much of it does not decompose quickly and accumulates.
- Small particles or chemicals (which may be toxic) are absorbed by plants and passed on to the animals that eat those plants. This continues up the food chain and the chemicals become more concentrated higher up the food chain. This is called **bioaccumulation**.
- Even accumulation of non-toxic materials can cause harm. When mineral nutrients from excess fertiliser accumulate in water they cause increased growth of algae in the water. This is called **eutrophication**. As the algae die they are decomposed by large numbers of bacteria that use up all the oxygen in the water – this kills the other living organisms in the water.

Tip

Sustainable means that the process can continue indefinitely without causing harm. Only closed-loop systems are sustainable.

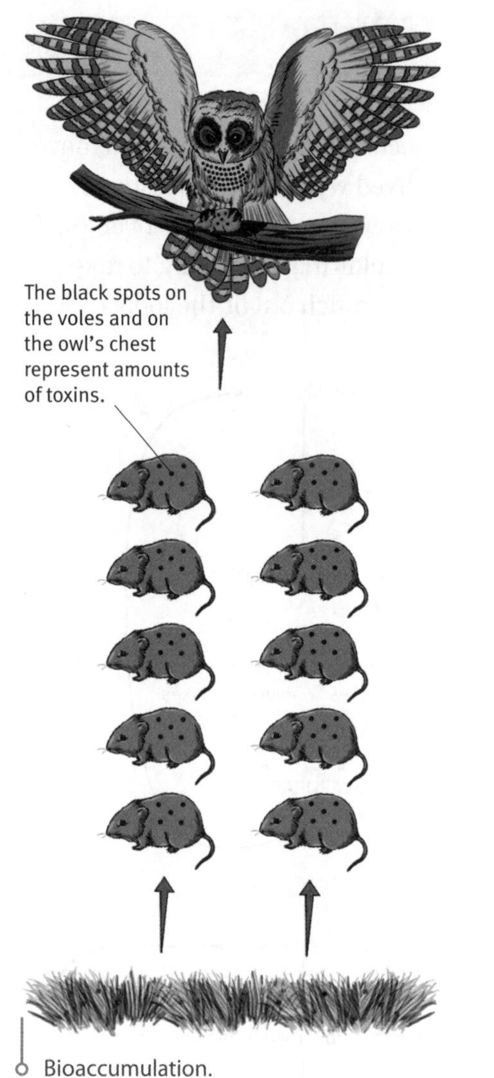

Bioaccumulation.

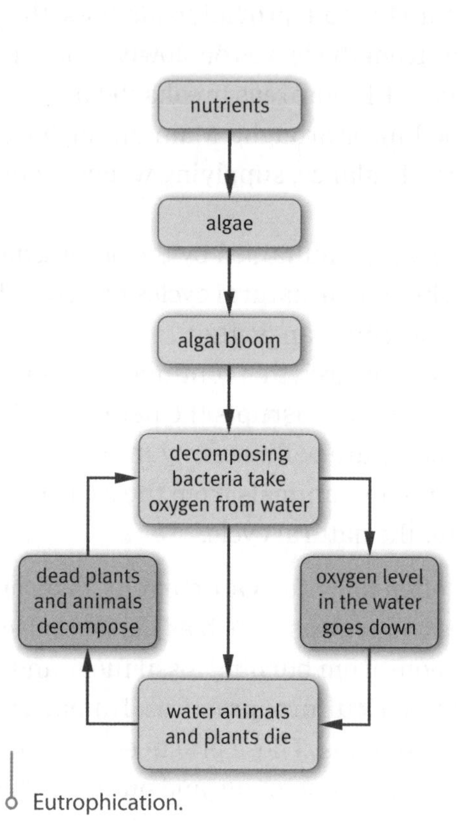

Eutrophication.

- Continuous cropping of food plants or timber removes biomass from the ecosystem. This means that chemicals are removed and cannot be recycled, so the natural ecosystem runs out of mineral nutrients and the natural cycles break down. Plants cannot grow without the minerals and the soil is exposed. With little dead organic matter soil can be eroded quickly – this causes **desertification**.
- Natural vegetation has a rich **biodiversity**. If natural vegetation is replaced by agricultural crops there is a great loss of biodiversity. When the crops are harvested the soil is exposed to erosion, which may lead to desertification. The lost soil can be washed into rivers causing **silting**.
- Use of fossil fuels such as crude oil alters the natural ecosystem because the fuel took millions of years to form from the decay of dead organisms. The energy collected from the Sun over millions of years when these organisms were alive is released quickly – this is known as fossil sunlight energy.

B7.4.19, B7.4.21–23 How can humans be more sustainable?

The rate at which we currently use natural resources is not sustainable. It can only be made sustainable if we use resources at a rate at which they can be replaced.

- Use of natural resources such as timber and fish must be controlled. This can be achieved by selective felling of specific species and mature trees, and by replanting. Fish must also be selectively harvested and strict **quotas** agreed. This requires international agreement. **Restocking** natural fisheries will also support increased harvesting.
- Energy sources must be sustainable. The Sun is the most sustainable source of energy. It supplies energy to natural ecosystems and to agriculture. It can be used to grow crops for fuel – **biofuels**. The Sun's energy can also be harvested through photoelectric cells to produce electricity and through solar heating systems. The Sun's energy also creates the wind and can be harvested through wind turbines.
- Sustainable use of energy and natural ecosystems is essential to avoid increased desertification, drought, flooding, and the loss of the natural ecosystems, which may lead to global catastrophe. However, it is hard for local populations to understand that their role is essential – they must not over-exploit their local ecosystems for short-term gain, such as the exploitation of tigers.

Use extra paper to answer these questions if you need to.

1 a What is meant by an open-loop system?
  b Describe one example of an open-loop system.
  c What happens to wastes in an open-loop system?
  d What is meant by a closed-loop system?
  e Describe one example of a closed-loop system.
  f What are used as reactants in a closed-loop system?
  g Which type of system do humans often create?

2 a What is meant by a sustainable process?
  b Explain why human systems are often not sustainable.
  c Why are ecosystems often not perfectly closed loops?
  d Name two possible losses from an ecosystem.
  e Explain why the input to an ecosystem should equal the losses.

3 a What is the source of energy for all ecosystems?
  b Describe how carbon is recycled in an ecosystem.
  c How does burning fossil sunlight energy alter the system?
  d Describe how nitrogen is recycled in an ecosystem.

4 a Explain the role of microorganisms in an ecosystem.
  b Name one type of decomposer.
  c Explain the role of digestive enzymes in decomposition.

5 a Explain what is meant by 'take–reuse–recycle'.
  b Why is a 'take–reuse–recycle' policy better than a 'take–make–dump' policy?
  c Explain why some plants apparently over-produce pollen.

6 a What is meant by deforestation?
  b Explain why deforestation can lead to flooding and drought.
  c Describe how deforestation can damage the soil.
  d What effect might this have on rivers and lakes?

7 a What is meant by biodegradable?
  b Name two substances that are not biodegradable.

8 a What is an ecosystem service?
  b Describe two ecosystem services.
  c What is meant by over-exploitation?

9 a How do we exploit the seas?
  b Suggest a possible consequence of over-exploiting the seas.

10 a Describe the effect of continuously cropping the same food plant from soil.
  b What effects do intensively farmed livestock have on the ecosystem?
  c Explain how intensive agriculture reduces biodiversity.
  d Explain why it is important to maintain biodiversity on a farm.
  e Why is it important to not use too much fertiliser?

11 a How long does crude oil take to form?
  b What is crude oil made from?
  c Where did the energy in crude oil originally come from?
  d Explain why using crude oil does not fit in with a closed-loop system.

12 a What is a quota?
  b Explain why fishing quotas are important.
  c Explain why international cooperation may be essential to maintain fish stocks.
  d What is meant by restocking?
  e Explain why only adult fish should be caught.

13 a Describe the techniques used to prevent deforestation.
  b What are nitrogen-fixing bacteria?
  c How can the action of nitrogen-fixing bacteria be maximised in agriculture?
  d How does crop rotation help to maintain soil stability?
  e Describe two techniques used in agriculture to overcome the problem of soil erosion.

14 a List three ways in which we make use of the Sun's energy.
  b Explain why using the Sun's energy is sustainable.

15 a Describe how a local human community might make use of the natural ecosystem sustainably.
  b Why might some local human communities start to over-exploit their natural ecosystem?
  c Give three reasons why it is important to conserve individual species such as the tiger.
  d Explain why natural ecosystems must be conserved.

H

16 a What is bioaccumulation?
  b Explain why organisms low down a food chain may be unharmed by toxic substances.
  c Explain why organisms higher up the food chain are more likely to be harmed by a toxic compound.

17 a Describe the process of eutrophication.
  b What is an algal bloom?
  c What process uses up oxygen in eutrophication?
  d Which organisms use up most oxygen during eutrophication?
  e What actually kills the fish in eutrophication?

1 The flow diagram shows a system in which resources from natural ecosystems are used for manufacturing and consumption.

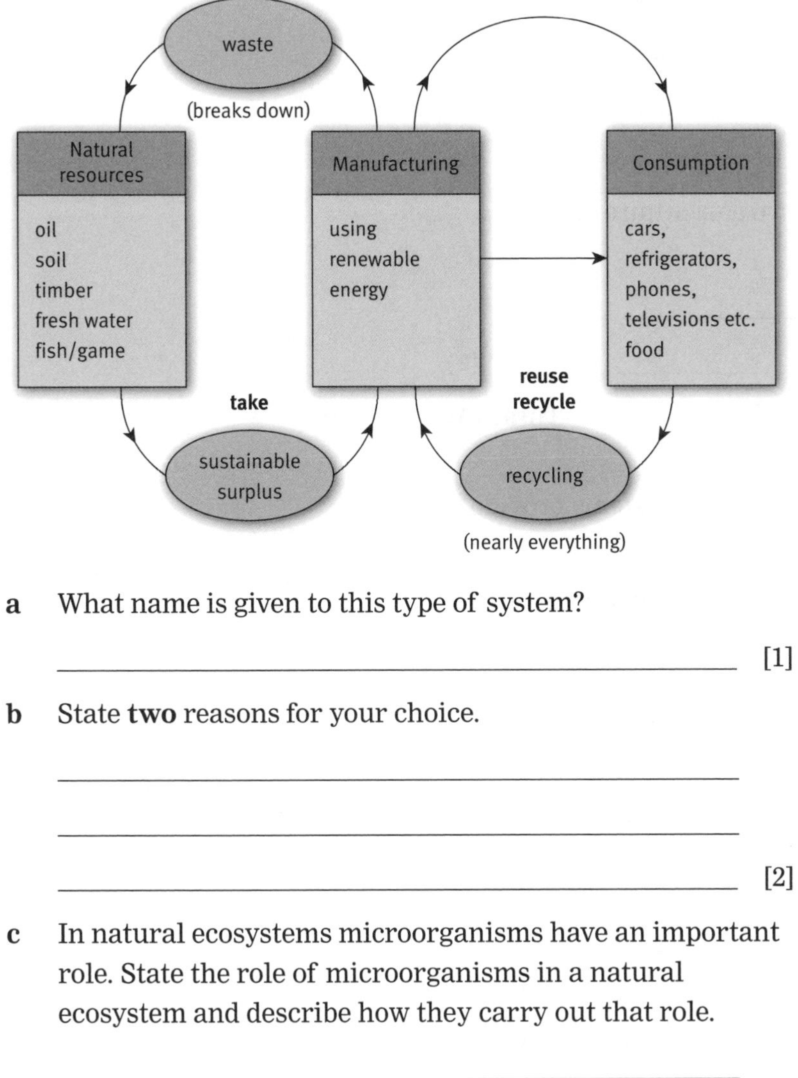

a What name is given to this type of system?

______________________________ [1]

b State **two** reasons for your choice.

____

____

______________________________ [2]

c In natural ecosystems microorganisms have an important role. State the role of microorganisms in a natural ecosystem and describe how they carry out that role.

____

____

____

______________________________ [3]

Total [6]

2 Describe and explain the effects of deforestation.

✎ The quality of written communication will be assessed in your answer to this question.

Write your answer on separate paper or in your exercise book.

Total [6]

3 a What is meant by intensive agriculture?

____

______________________________ [1]

b Intensive agriculture is an open-loop system. One input is the addition of fertiliser.

Explain why fertiliser must be added.

____

____

______________________________ [2]

c List **two other** inputs to intensive agriculture such as growing almonds in California.

____

______________________________ [2]

Total [5]

4 Ecosystem services are ways in which living systems provide for human needs.

a Explain the role of bees in providing for human needs.

____

____

______________________________ [2]

b Describe how natural vegetation protects soil from erosion.

____

____

______________________________ [2]

c Describe how forests can affect the provision of clean water.

____

____

____

______________________________ [3]

Total [7]

H **5** Eutrophication can result from the addition of too much fertiliser. Describe what happens during eutrophication and the effects it can have on the waterways.

____

____

____

____

____

Total [4]

1 Complete the sentences by crossing out the incorrect word.

Given suitable conditions bacteria reproduce **quickly** / **slowly**.

Bacteria can be used to manufacture **enzymes** / **stem cells**.

Fungi can be grown in **kettles** / **fermenters**.

A **vectra** / **vector** is used to insert genes into bacterial cells.

Nanotechnology uses particles that are **100** / **10 000** nanometres long.

2 Place the following steps in the correct order in the boxes below to describe the process of genetic modification.

A Select the modified individuals.

B Put the gene into a suitable vector.

C Isolate and replicate the required gene.

D Use the vector to insert the gene into a new cell.

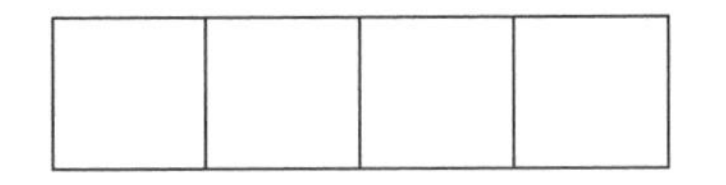

3 Explain the following statements.

a Cheeses made using chymosin can be eaten by vegetarians.

___

b Stem cells can develop into a range of cell types.

___

c Bacteria can manufacture human insulin.

___

d Many washing powders are called biological washing powders.

___

4 Put a ring around **T** or **F** to show whether each statement is **true** or **false**.

Statement		
Heart valves can be replaced.	**T**	**F**
New skin can be grown using tissue culture.	**T**	**F**
Fungi can be used to make medicines.	**T**	**F**
Quorn is protein made from bacteria.	**T**	**F**
Stem cells can only be found in plants.	**T**	**F**
Plants can be genetically modified to grow human tissue.	**T**	**F**
Bone marrow contains stem cells.	**T**	**F**
Biofuels produce no carbon dioxide.	**T**	**F**

5 Complete the word grid. All the words are used in this topic.

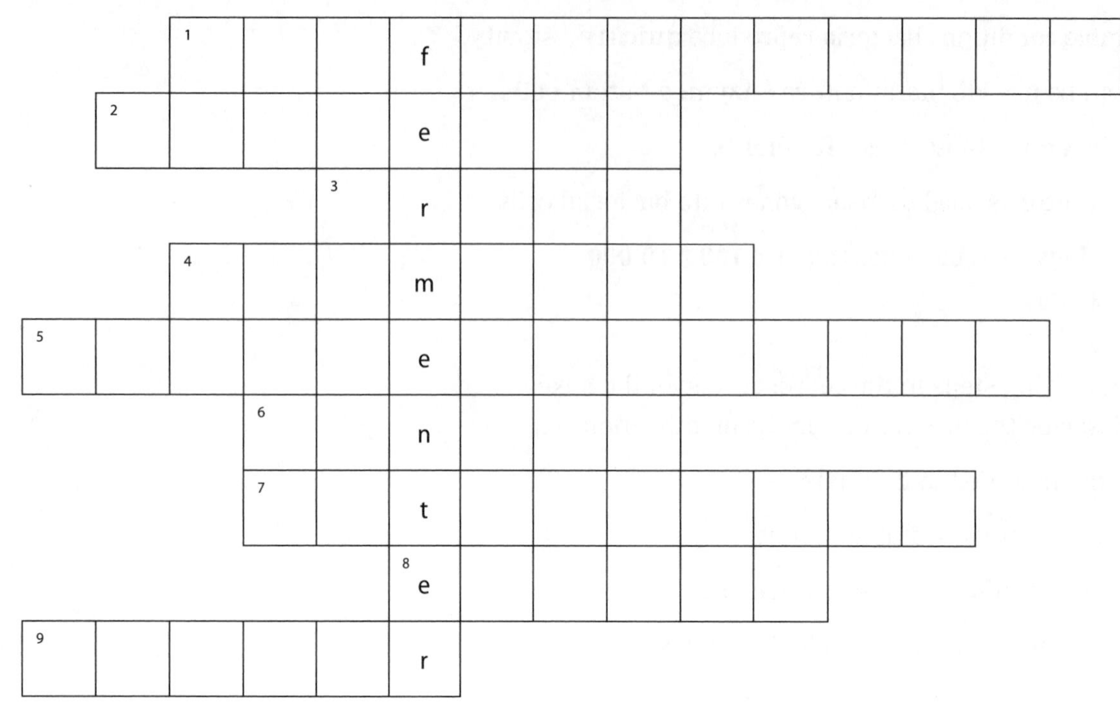

Clues

1 cells do this to become specialised
2 single-celled organisms that contain plasmids
3 the name for a genetic marker
4 another name for rennet
5 technology using tiny particles
6 an organism used to grow single-cell protein
7 a chemical made by fungi used in medicine
8 a chemical used in biological washing powder
9 something used to insert genes into a bacterium

H **6** A gene probe for cystic fibrosis contains the sequence AAG CGC.

The list shows sections of DNA from four people. Which of them has cystic fibrosis?

Sally	AAG CGC
Peter	ACG TGA
David	TTC GCG
Wendy	GGA TCT

____

B7.5.1 The role of bacteria

There is a huge range of different microorganisms. Because they are living organisms they produce a wide range of chemicals as part of normal life. These include **enzymes** and **antibiotics** that we find very useful.

We can grow microorganisms in conditions that ensure they produce large amounts of these chemicals. Bacteria are particularly useful as they have certain features:

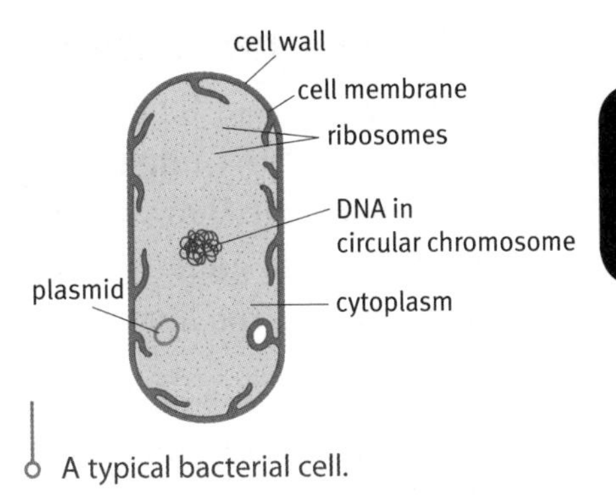

A typical bacterial cell.

Feature	Benefit
rapid reproduction	The population grows rapidly and can produce large amounts of the useful products quickly.
presence of plasmids	Plasmids are genetic material that can be used to introduce new genes – this means that we can make the bacteria produce what we want.
simple biochemistry	The nutrients and conditions needed for growth are simple and can be supplied easily.
ability to make complex molecules	Despite their simple biochemistry bacteria can still make complex molecules including enzymes, antibiotics, food additives, and hormones.
lack of ethical concerns in their culture	There are no animal welfare issues – the bacteria are often kept in their ideal conditions and are removed to be reused at the end of the process.

B7.5.2 Fermentation

Microorganisms include bacteria and fungi. They can be grown in industrial tanks called **fermenters**. Many microorganisms are grown in batches under ideal conditions that are carefully monitored and controlled. Excess heat and toxic waste products are removed.

- Most antibiotics are produced by fungi. The **fungus** *Penicillium* makes penicillin, which kills bacteria and can be used to treat infections.
- Fungi can be used to make a protein called **single-cell protein**. This can be extracted and processed to eat. The product is called Quorn.
- Bacteria and fungi can be used to make enzymes used in food manufacture. For example, the enzyme **chymosin** or **rennet** is used to make cheese.
- Bacterial enzymes can also be used in washing powders to help digest stains made by fats, proteins, or carbohydrates.
- **Biofuels** can be made using enzymes from microorganisms. Ethanol is made by fermenting sugar and a newly developed enzyme (lignocellulase) can digest woody tissue.

B7.5.3–5 Genetic modification

Bacteria hold some of their genetic material in **plasmids**. These are small rings of DNA that are easy to modify. Inserting a gene for a human protein into a plasmid and placing the plasmid in a bacterium is called **genetic modification**. The bacterium uses the information in the human gene to make the human protein.

The main steps in genetic modification are:

1 isolating and replicating the required gene (e.g., the gene coding for **insulin**)
2 placing the gene into a **vector** (a plasmid or a **virus**)
3 using the vector to insert the gene into a new cell (plasmids are small and can move in and out of bacterial cells easily)
4 selecting the modified individuals.

Examples of successful genetic modification include:

- using bacteria to manufacture medicinal drugs such as insulin (the hormone used to treat diabetes)
- making crop plants **resistant** to **herbicides** so that crops can be sprayed to kill weeds.

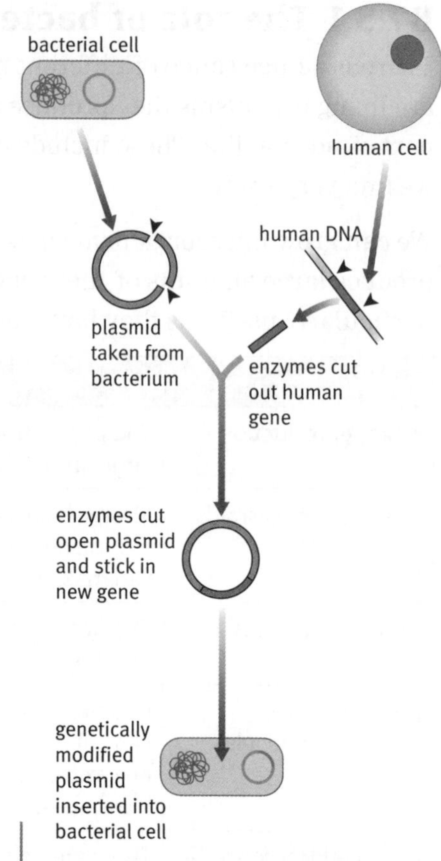

Genetic modification (not to scale).

H B7.5.6 DNA technology

Genetic testing involves testing a person's blood to see if it contains a particular gene – perhaps a gene for an inherited disorder.

DNA consists of two strands that bind to each other using **complementary** sequences of bases. This means that specific sequences can be identified using known sections of DNA as markers (**gene probes**). It is a complex process:

1 Some DNA is extracted from white blood cells.
2 The DNA is isolated and cut up using enzymes.
3 The DNA is heated to split the complementary strands creating single-stranded DNA with its base sequence exposed.
4 A gene probe is made. This is a length of single-stranded DNA that carries a complementary base sequence to the gene being looked for. A fluorescent marker is attached to it.
5 The gene probe is mixed with the DNA sample and will bind (using its complementary base sequence) to the gene that is being looked for.
6 The DNA is separated using **gel electrophoresis**. This uses the size of the DNA fragments to separate out the fragments that have probes attached from the remaining single-stranded DNA.
7 UV light is shone on the sample to locate the DNA fragments that have the fluorescent marker and probe attached.

This technique can be used to identify people with faulty alleles such as an inherited disease or to locate a gene on a chromosome.

Tip

Remember that insulin made by genetic modification is human insulin from a human gene, so there are no ethical or medical issues about its use.

Tip

Remember that DNA technology relies on complementary (matching) base pairs.

B7.5.7–8 Nanotechnology

Nanotechnology is technology using tiny particles that are no larger than some molecules. The particles used are no larger than 100 **nanometres**.

Examples of nanotechnology include:

- silver particles embedded in plastic food wrapping help preserve the food; the silver particles inhibit bacterial growth
- antibodies in food wrapping react with bacteria and inhibit their growth
- nanoparticles in food wrapping respond to oxygen levels, indicating that the wrapping may be damaged
- nanoparticles respond to ripening fruit to indicate the condition of the food.

B7.5.9 Stem cell technology

Stem cells are cells that have not specialised or **differentiated**. This means that they can divide and specialise to become a range of different types of cell.

Stem cells can be used in medical treatments to grow new **tissues** and there is potential to grow whole new **organs**.

Leukaemia can be treated by transplanting bone marrow, which contains stem cells that are able to divide and differentiate into a range of blood cell types. The new marrow enables the patient to make normal blood.

Tissue culture can grow new tissues in the laboratory – for example, skin can be grown for grafts to treat burns.

It may soon be possible to treat a range of problems using stem cells including spinal cord injuries that cause paralysis, diabetes, and Alzheimer's disease.

B7.5.10 Biomedical engineering

The heart is a good example of an essential organ for which **biomedical engineering** can restore proper functioning if there is a fault. There are two faults that can be corrected relatively easily:

- The rhythm of the heartbeat is controlled by an area of the heart called the **pacemaker**, which creates electrical signals to stimulate the heart muscle. If this is faulty the electrical signals can be created by an artificial pacemaker embedded under the muscle of the chest.
- Faulty **heart valves** that ensure the correct flow of blood through the heart can be replaced by artificial versions or by real valves from a donor.

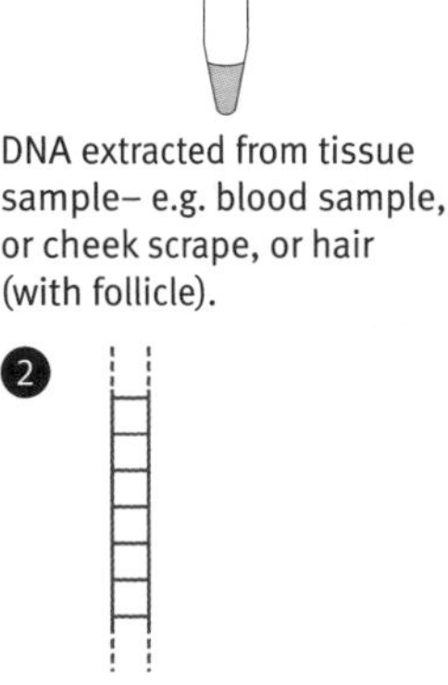

1 DNA extracted from tissue sample– e.g. blood sample, or cheek scrape, or hair (with follicle).

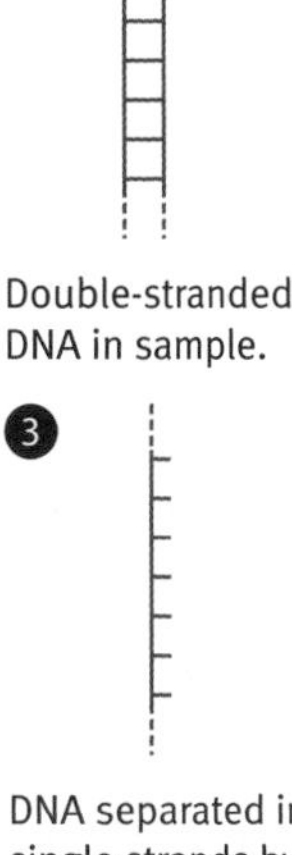

2 Double-stranded DNA in sample.

3 DNA separated into single strands by gentle heating.

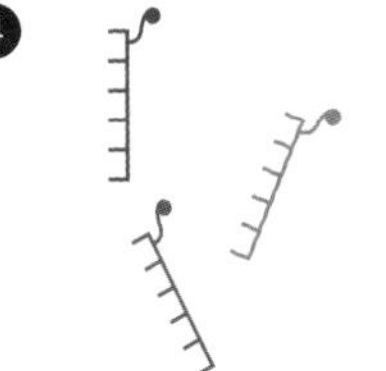

4 Short sections of DNA with fluorescent markers are added. The sequence of the short section is complementary to a target section in the original DNA.

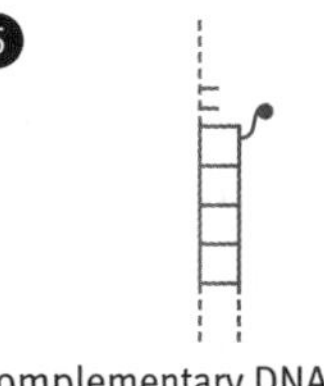

5 Complementary DNA binds if it matches the target sequence.

Genetic profiling and the use of gene probes.

Tip

A nanometre is one-millionth of a millimetre. You could fit 10 000 nanoparticles along a 1-mm length.

Tip

Stem cells are found in a number of tissues including bone marrow, umbilical cords, and skin.

Use extra paper to answer these questions if you need to.

1 a What are microorganisms?
 b How are bacteria different from human cells?
 c Explain why bacteria may be described as living factories.
 d List five features of bacteria that mean they are ideal for industrial manufacturing.
 e How are fungi different from bacteria?

2 a Explain what is meant by fermentation.
 b List two products made by fermentation.
 c What conditions need to be controlled in a fermenter?

3 a What is single-cell protein?
 b What type of organism makes single-cell protein?
 c How can single-cell protein be used?

4 a What is the source of rennet?
 b How is rennet used in the food industry?
 c Name a source of chymosin, other than rennet.
 d Explain why cheese made using this chymosin is suitable for vegetarians.

5 a What are biofuels?
 b How is ethanol made from sugar?
 c Why might someone argue that crops should not be grown specially to make biofuels?
 d Name the enzyme that has been developed to digest woody tissue.

6 a What is a biological washing powder?
 b What types of stain can they work best on?

7 a What is meant by 'genetic modification?
 b Name the rings of DNA found in bacteria.
 c Why does a bacterium need a human gene in order to make a human protein?
 d Name the human protein produced by bacteria that is used to treat diabetes.
 e What is a vector?
 f Explain why plasmids are good vectors.
 g Describe how the human insulin gene can be inserted into a bacterium.

8 a Give one example of genetic modification in plants.
 b Explain the advantages of this genetic modification.
 c Give three arguments against genetic modification.
 d Offer an alternative argument for each point you have made.

9 a What does the prefix 'nano' mean?
 b How big are the particles used in nanotechnology?
 c Describe three examples of how nanoparticles can be used in food packaging.
 d How many nanometres are there in a metre?
 e Name a metal commonly used in nanotechnology.

10 a What is meant by a differentiated cell?
 b Why are cells differentiated?
 c What is a stem cell?
 d Where in the body do stem cells occur?

11 a What is meant by tissue culture?
 b Why might skin need to be replaced in a medical procedure?
 c Describe how stem cells can be used to treat leukaemia.
 d List three possible future uses for stem cells in medical treatments.
 e Explain why a donor is needed when using stem cells in medicine.
 f Explain why the donor must be matched to the recipient.

12 a What is biomedical engineering?
 b Name two parts of the heart that can be repaired using biomedical engineering.
 c What is the role of a pacemaker in the heart?
 d Why might the pacemaker need to be replaced?
 e What is the role of valves in the heart?
 f Why might the valves need to be replaced?
 g Name two possible sources of new valves.
 h Describe three properties of the materials used to make artificial valves and pacemakers.

H

13 a Briefly describe the structure of DNA.
 b Describe two features of DNA structure that allow genetic testing to occur.
 c What tissue is usually sampled to extract DNA for genetic tests?
 d Explain why red blood cells cannot be used for genetic testing.

14 a What is a gene probe?
 b Why must DNA be heated before it is mixed with a gene probe?
 c Why are gene probes attached to fluorescent markers?
 d How are fragments of DNA with probes attached separated from other fragments of DNA?
 e Name two genetic disorders that could be tested for using genetic testing.

1 a Name two organisms that can be grown in a fermenter.

______________________ [2]

b Name two products that can be made in a fermenter.

______________________ [2]

Total [4]

2 A student decided to investigate the effect of changing temperature on the rate of fermentation.

a i Write a hypothesis about the effect of temperature on the rate of fermentation and predict the results you might expect. [2]

____

____

ii Explain this hypothesis and prediction. [2]

____

She was given a suspension of microorganisms in glucose solution and the apparatus shown below right. Fermentation releases carbon dioxide, which turns the hydrogencarbonate indicator from pink to yellow.

b i Suggest what type of microorganism is used. [1]

____

ii Identify two additional pieces of apparatus that would be essential for the student to carry out this investigation. [2]

____

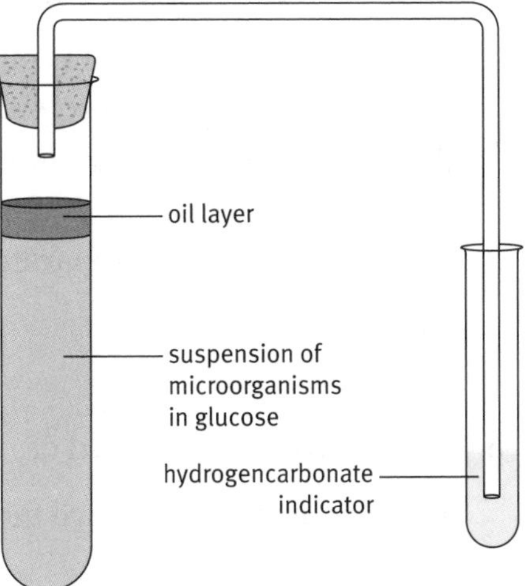

The student divided the suspension of microorganisms into five portions and heated each portion to a different temperature. She measured how long it took for the indicator solution to turn yellow. She recorded the following results.

Temperature (°C)	Time (s)	Rate (s^{-1}) (Rate = $\frac{1}{\text{time}}$)
10	432	0.002
20	312	0.003
30	148	0.007
40	74	0.014
50	78	0.013

c i State whether these results back up your hypothesis and justify your statement. [3]

____

____

ii Suggest how the student could modify her investigation to improve your confidence in the results. [1]

____

Total [11]

3 Suggest some advantages of:

a producing insulin using genetically modified bacteria

____

______________________________ [2]

b producing cheese using chymosin

____

______________________________ [2]

c making single-cell proteins from fungi

____

______________________________ [2]

d using silver nanoparticles in food wrapping.

____

______________________________ [2]

Total [8]

Going for the highest grades

4 Describe how a bacterium could be genetically modified to manufacture a human protein such as insulin.

The quality of written communication will be assessed in your answer to this question.

Write your answer on separate paper or in your exercise book.

Total [6]

H 5 A gene probe is composed of two parts: DNA and a marker.

a How many strands are there in the DNA of a gene probe?

______________________________ [1]

b Explain how the marker is made visible

______________________________ [1]

c Explain how a gene probe is used.

____

______________________________ [2]

Data: their importance and limitations

Temperature (°C)	Volume of gas collected per minute (cm³)
10	0
20	1
30	25
50	1
80	0

Data: their importance and limitations

1 A student decided to improve his fitness. He carried out an exercise programme and monitored his fitness by measuring his heart rate.

Place a tick next to the factors he should consider to ensure he collects good data.

He should take repeat readings	☐
He should measure his muscle strength	☐
A fit person will have a lower heart rate	☐
The time of day when his pulse is counted	☐
His diet	☐
He should calculate a mean	☐

2 Complete the following statements. Choose from the words below to fill in the gaps.

incorrect	repeatable	average	true	verify	false

If all the readings for a particular value are similar the data is described as ______________________.

If a measurement lies well outside the range within which the other measurements lie then it may be ________________.

We can never be certain that a measurement gives us the ______________ value of the quantity being measured.

Data is used to ________________ proposed explanations.

Data: their importance and limitations

1 Lilia is investigating how temperature affects the rate of photosynthesis.

She sets up the apparatus below opposite.

She draws a table for her results.

Temperature (°C)	Number of bubbles produced in 1 minute
10	
20	
30	
40	
50	

water
pondweed
paperclip 'weight'

a Lilia collects a set of results.

Describe how she could find out if the results are repeatable.

______________________________ [2]

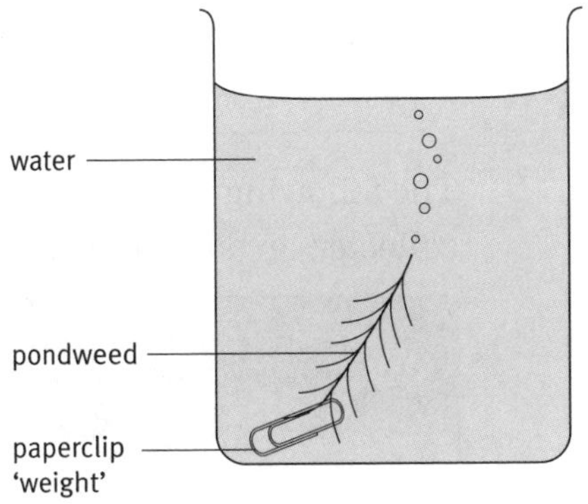

H **b** Lilia's teacher asks if her results are reproducible. Suggest how Lilia could find out.

______________________________ [2]

Total [4]

2 David carried out a programme to improve his fitness. The table shows his heart rate before and after exercise in four successive weeks.

Time since start of fitness programme (days)	Resting heart rate (bpm)	Heart rate after exercise (bpm)
0	73	147
7	71	145
14	69	143
21	65	148
28	62	138

a **i** The data recorded are mean values calculated after David carried out the fitness test three times. Explain why three readings were taken and a mean calculated.

____

____ [2]

ii Suggest one precaution David had to take to make his results valid.

____ [1]

b The following table shows the raw data that David collected on day 7 and day 21.

Day	Heart rate after exercise (bpm)			
	1st count	2nd count	3rd count	Mean
7	143	147	145	145
21	139	143	162	148

Using both tables of data:

i Discuss the evidence that David's fitness has improved.

____

____

____ [3]

ii Evaluate his data.

____

____

____ [3]

Total [9]

Cause–effect explanations

7
Exactly. Let's find out more about the samples.
Well, there were 44 volunteers altogether. That's a reasonable size, but I'd be more confident in the conclusions if the sample size had been even bigger.
8
The investigation was a randomised, controlled trial. That means that the volunteers were randomly put into either the white- or the dark-chocolate group.
9
All factors were the same, except the type of chocolate. Neither group did more exercise than the other, so this couldn't have caused the blood-pressure change.
10
In random groups, other factors are equally likely in both groups. The average age, weight, and waist size for each group were almost the same in each group – the scientists checked.
11
So there seems to be good evidence that the factor caused the outcome.
H
But I'd be happier if there was a **mechanism** that linked the dark chocolate to the lowered blood pressure.
12
H
There is! The dark-chocolate volunteers had extra nitrogen monoxide in their blood. Scientists think that flavanol compounds in dark chocolate made the amount of nitrogen monoxide increase. Nitrogen monoxide helps make blood vessels wider. And this can reduce blood pressure.

Cause–effect explanations

Cause–effect explanations

1 The graph shows the relationship between body mass index (BMI) and time taken to recover after exercise.

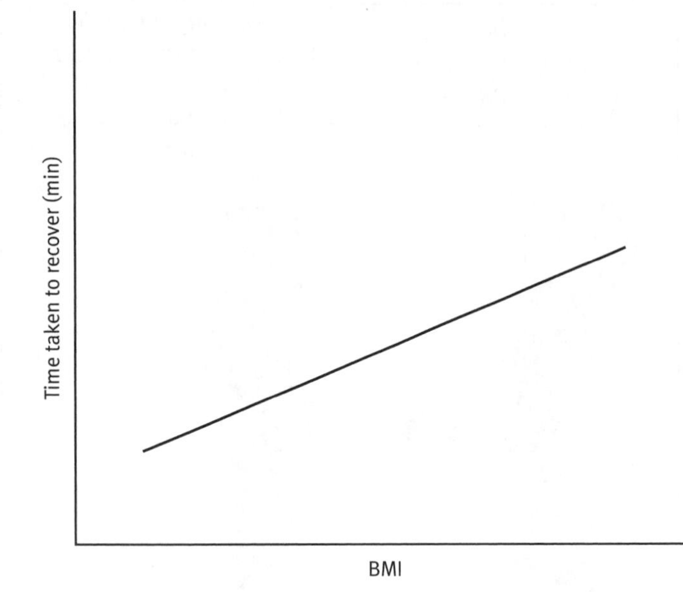

Which of the following statements are true?

a	As BMI increases, recovery time decreases.	**T**	**F**
b	A long recovery time causes a high BMI.	**T**	**F**
c	There is a correlation between BMI and recovery time.	**T**	**F**
d	As BMI increases, recovery time also increases.	**T**	**F**
e	A high BMI causes an increase in recovery time.	**T**	**F**
f	A high BMI causes a higher heart rate.	**T**	**F**

2 Solve the anagrams.
Match each anagram answer to a clue.

Clues
A To investigate whether a factor affects an outcome, we compare samples. The ____________________ the samples, the more confident we can be in the conclusions.
B If an outcome variable increases as an input variable increases, there is a ____________________ between the variables.
C Eating food high in saturated fats increases your ____________________ of having a heart attack.
D When investigating the relationship between a factor and an outcome, it is vital to ____________________ all the factors that might affect the outcome.

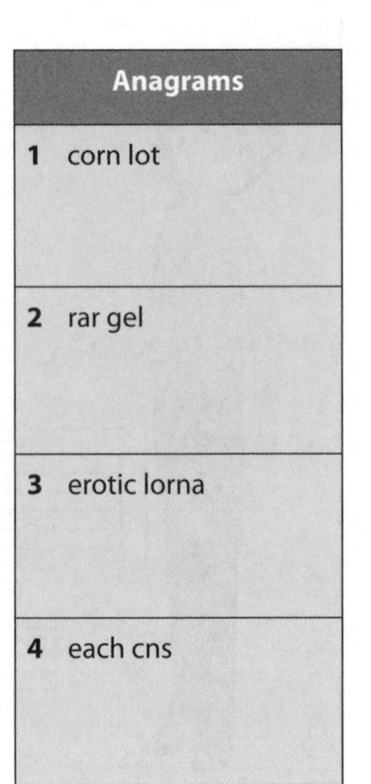

Anagrams
1 corn lot
2 rar gel
3 erotic lorna
4 each cns

Cause–effect explanations

1 A scientist plans to investigate the effect of carbon dioxide concentration on the flavour of strawberries.

A strawberry has an intense flavour if it contains a relatively high mass of ester compounds.

The scientist grows three groups of strawberries. He adds different amounts of extra carbon dioxide to the air around two of the groups.

Each group of strawberries receives the same amount of water and is grown at the same temperature.

When the strawberries are ripe, the scientist picks them. He measures the masses of ester compounds in the strawberries of each group.

a Identify the outcome variables.

____ [1]

b **i** Identify two factors (variables) that the scientist controls.

____ [1]

ii Explain why he controls these factors.

____ [1]

c The table gives some results of the investigation.

Concentration of carbon dioxide in air (ppm)	Concentration of ester compound A in strawberries (ng/g)	Concentration of ester compound B in strawberries (ng/g)	Concentration of ester compound C in strawberries (ng/g)
350	651	531	53
650	815	637	54
950	963	715	53

i Describe two correlations shown by the data in the table.

____

____

____ [2]

ii Four students discuss the results in the table.

Tim The correlations show that the factor causes the outcomes.

Kezi The factor might cause the outcomes, but we cannot be sure.

Sahira The factor and the outcomes might both be caused by some other factor.

Ashfaq There is no correlation between the factor and one of the outcomes. This shows that the factor cannot cause the other outcomes.

Give the names of the two students whose statements are correct.

__________________ and ________________ [2]

H **d** The scientist suggested *how* extra carbon dioxide gas in the air may help strawberry plants to synthesise greater amounts of ester compounds.

Explain why the scientist's suggestion may make other scientists more willing to accept that the factor causes the outcomes shown in the table in part **c**.

____

____ [1]

Total [8]

2 A group of scientists investigated the effect of taking aspirin on the chance of getting bowel cancer.

The scientists asked people with Lynch syndrome to help them with their research. People with Lynch syndrome are more likely to get bowel cancer than other people because their bodies struggle to detect and repair damaged DNA.

a Write an O next to the **two** statements that are part of the outcome.

Write an F next to each of the **two** statements that are factors that may affect the outcome.

Of the people who took aspirin, 10 got cancer. __

For 2 years, 258 people took aspirin every day. __

For 2 years, 250 people took aspirin placebo tablets (tablets that look like aspirin but have no aspirin in them) every day. __

Of the people who took aspirin placebo tablets, 23 got cancer. __ [2]

b Use the statements in part **a** to help you answer this question.

Tick the boxes next to all the statements below that are true.

There is a correlation between the factors and the outcome. ☐

Taking aspirin reduces the chance of getting bowel cancer. ☐

Taking aspirin increases the chance of getting bowel cancer. ☐

Taking aspirin means a person will not get bowel cancer. ☐ [2]

c Suggest why the groups of people in the study (aspirin or placebo) were chosen randomly.

____

____ [1]

H **d** The scientists suggested a mechanism to explain how aspirin prevents bowel cancer. They think that aspirin may kill stem cells in the bowel that cannot repair damaged DNA.

Tick the box next to the two statements below that are true.

The suggested mechanism proves that taking aspirin causes a reduced chance of bowel cancer. ☐

The suggested mechanism increases confidence that taking aspirin causes a reduced chance of bowel cancer. ☐

The suggested mechanism decreases confidence that taking aspirin causes a reduced chance of bowel cancer. ☐

It is possible that the decreased chance of getting cancer after taking aspirin for 2 years is caused by some other factor. ☐ [2]

e Another group of scientists researched the effects of aspirin on several different cancers.

They studied 12 500 patients for 20 years. They found that the risk of cancer death was reduced by 20% for patients taking aspirin.

i Suggest two factors that might affect the risk of getting cancer, other than taking aspirin.

____

____ [2]

ii Identify one aspect of the study that increases confidence in its conclusions.

____ [1]

Total [10]

3 Bob observed that some pondweed in his fish tank bubbled when sunlight shone on the tank. He thought that this was because the pondweed was photosynthesising. He decided to test this idea by placing the tank in the dark and shining a light with a dimmer switch on the tank. He expected there to be more bubbles when the light was brighter.

a What was Bob's observation?

____ [1]

b What was Bob's prediction?

____ [1]

c Suggest an explanation for Bob's prediction.

____

____ [2]

d Bob's results are show below.

Brightness of lamp	Number of bubbles per minute
off	0
dim	0
medium	0
bright	18

i Does Bob's data match his prediction?

____ [1]

ii Suggest an alternative explanation to account for Bob's results.

____

____ [2]

iii How should Bob modify his test?

____ [1]

Total [8]

Developing scientific explanations

Temperature (°C)	Volume of gas collected per minute (cm^3)
10	0
20	1
30	25
50	1
80	0

Developing scientific explanations

1 The statements below describe how a group of scientists tested a hypothesis about the evolution of banded snails between 2009 and 2011. Banded snails have a variety of different shell colours.

Write the letter of each statement in a box on the flow chart. There is one letter for each box.

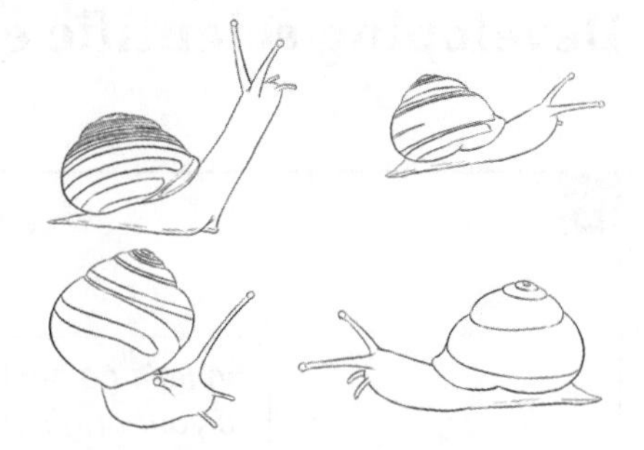

A Snails have evolved since the 1970s because the climate has got hotter, and a light-coloured shell may protect a snail from overheating.

B In 2009 and 2010, more than 6000 people in 15 countries counted banded snails of different shell colours. A scientist found records of snail shell colours from the 1970s.

C Between 1970 and 2010 there was an increase in the percentage of snails with one dark spiral band around their shells. The percentage of snails with light-coloured shells had not changed.

D A scientist observed banded snails with different shell colours. The percentage of snails of each shell colour seemed to have changed since the 1970s.

E The percentage of snails with light-coloured shells will be greater in 2010 than in 1970.

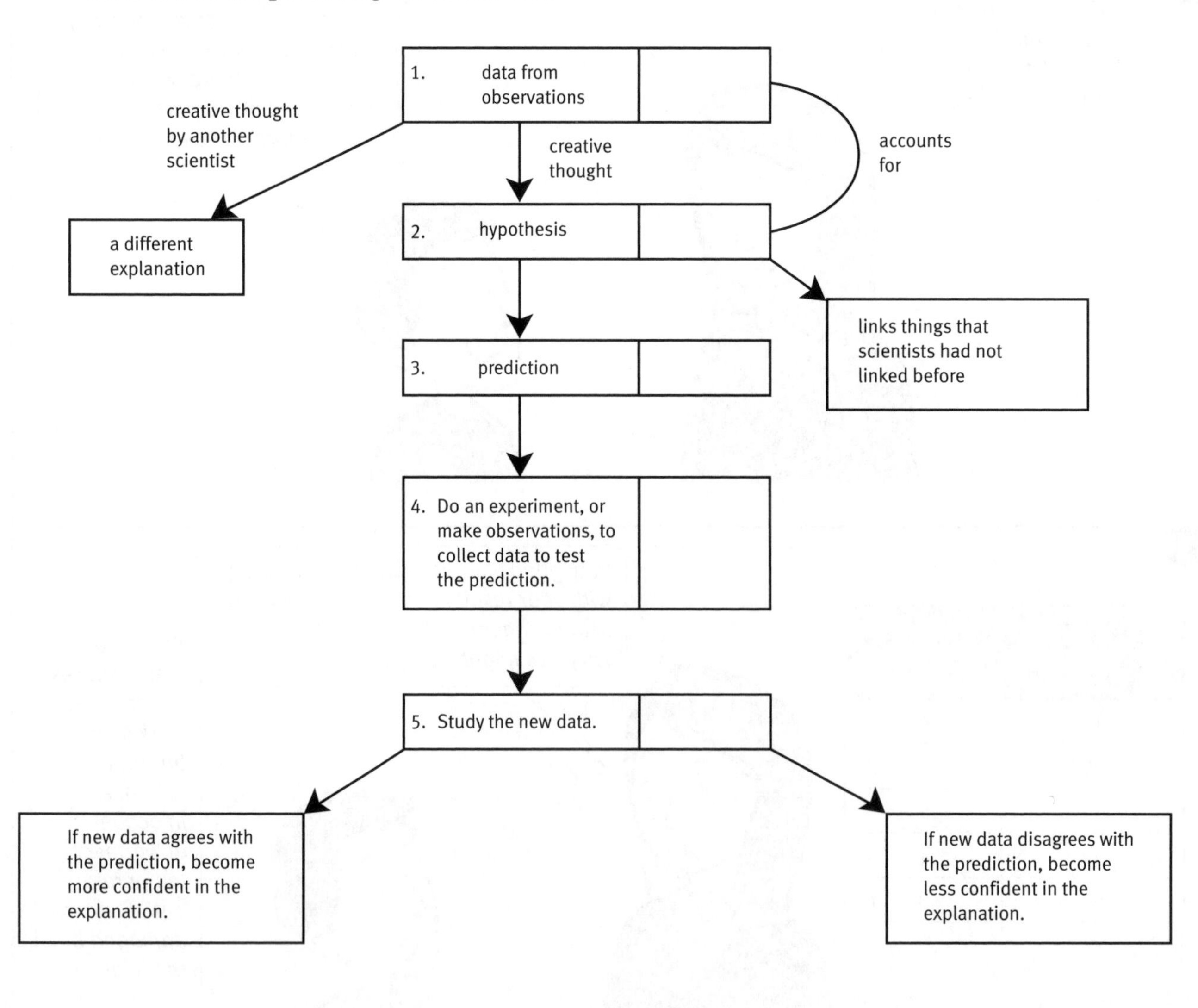

2 The statements below describe how scientists developed an explanation about the antibacterial substances in crocodile blood.

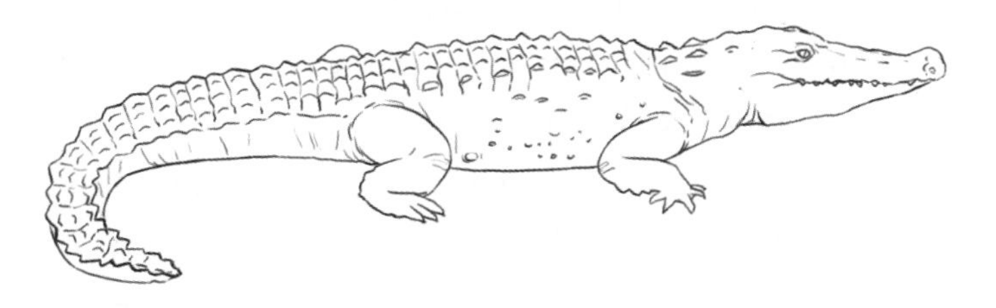

Write the letter of each statement in a box on the flow chart.
There is one letter for each box.
One letter has already been filled in.

A A scientist noticed that serious crocodile injuries heal quickly without getting infected.

B Crocodile blood will destroy microorganisms such as *E. coli*.

C The areas around the crocodile-blood-soaked discs had no *E. coli* bacteria.

D The scientist soaked paper discs in crocodile-blood extract. He placed these on agar plates with *E. coli* bacteria.

E Crocodile blood contains substances that destroy bacteria.

F After an injury, crocodiles wallow in mud that has antibacterial properties.

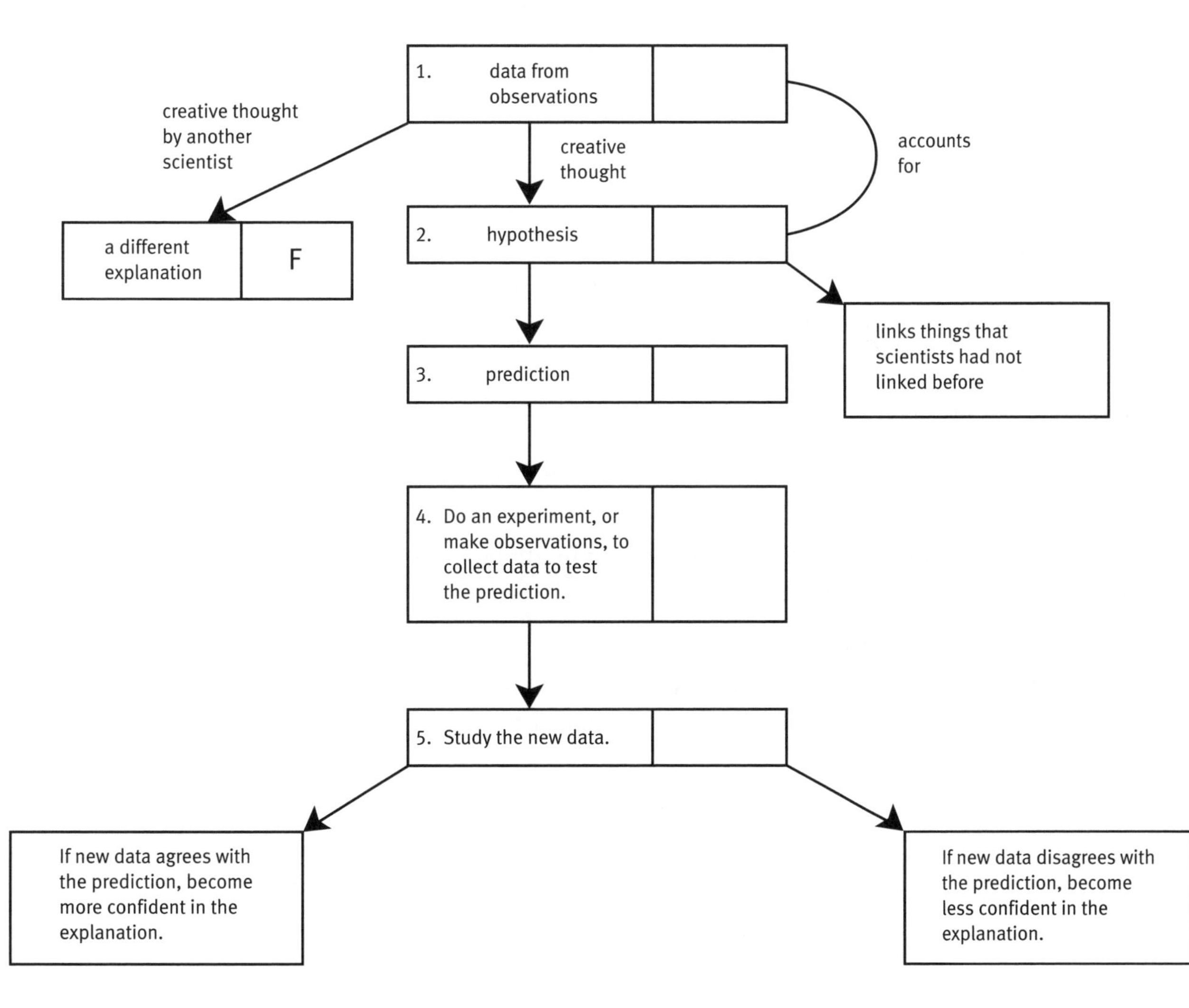

3 The statements below describe how a group of scientists tested a hypothesis about the size of a part of the brain in London taxi drivers.

Write the letter of each statement in a box on the flow chart. There is one letter for each box.

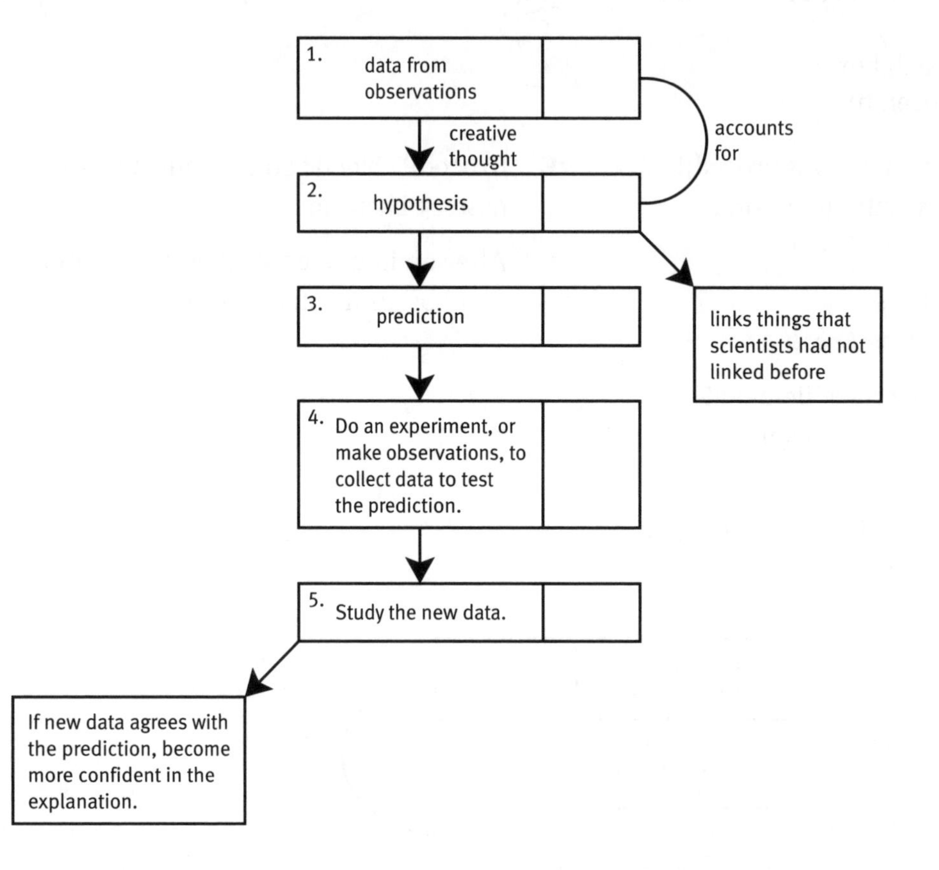

A London taxi drivers have an excellent knowledge of all the street names in a 280-km^2 area of London.

B We did MRI scans of 16 taxi drivers' brains, and of 50 other people's brains.

C The back of the hippocampus part of the brain was bigger in taxi drivers than in other people.

D The part of the brain that deals with navigation might be bigger in taxi drivers than in other people.

E If we the measure different brain parts in taxi drivers, and compare them to those of other people, there will be size differences in the part of the brain that deals with navigation.

4 Place the following statements in the correct order when trying to develop a scientific explanation. Write the numbers in the final column to show the sequence.

Step	Statement	Correct sequence
A	Compare the results with your predictions.	
B	Try to explain the new phenomenon.	
C	Retest your explanation.	
D	Modify your explanation if necessary.	
E	Observe a new phenomenon.	
F	Make a prediction based on your explanation.	
G	Test your explanation with an investigation.	

5 Explain why:

a If there is a new explanation, predictions are made based on the new explanation.

____

____

b Scientific explanations are tested by comparing predictions with data from experiments.

____

____

c Scientists look for an agreement between their prediction and their observations.

____

____

d If the prediction and the observations do not agree, a scientist will often discard one or the other.

____

____

Developing scientific explanations

1 In Kenya, elephants sometimes go onto farmland and ruin maize crops. Some farmers kill these elephants.

Scientists wanted to explain why elephants go onto farmland. They could then predict when elephants were likely to go onto farmland.
The scientists could then tell farmers when to make the most effort to guard their crops from elephants.

The scientists studied seven elephants.

They found out where the elephants went and what they ate.

The statements below describe the scientists' work.

A	The scientists thought that elephants might go onto farmland in the dry season only.	
B	Satellite tracking showed that six elephants spent all their time in the lowlands.	
C	Tail-hair analysis from one elephant showed that he ate grass from the lowlands in wet seasons and maize from farmland in dry seasons.	
D	In the dry season, there is not enough grass for elephants to eat. Some elephants get the food they need from shrubs and trees. But if there are not enough shrubs and trees, elephants take maize from farmland to eat.	
E	Satellite tracking showed that one elephant spent the wet season in the lowlands and the dry season in a forest near farmland.	
F	Tail-hair analysis of six elephants showed that they ate trees and shrubs in the dry season. In the wet season they ate grass.	
G	Fewer elephants will die if scientists find out when they are most likely to eat maize from farms.	

Some of these statements are **data** and one is a possible **explanation**.
Write a **D** next to the **four** statements that are data.
Write an **E** next to the **one** statement that is an explanation.

Total [5]

2 Read the article in the box.

It is about research reported in a scientific journal in 2011.

In a short visit to a small island, scientists identified more species of fanged frogs than on a bigger island nearby. They did not find any frogs from another group – the *Platymantis* group.

The scientists thought about what they had seen. They suggested that the fanged frogs had evolved more on the small island because there was no competition from *Platymantis* frogs.

The scientists expected that every type of habitat on the small island would have its own species of fanged frogs. They went to many of these habitats, and identified the fanged frog species in each one. Many habitats had their own species of fanged frogs. The frog species had different sizes, different webbing on their feet, and raised their young differently.

Explain how the description of the scientists' work in the box illustrates how scientific explanations develop.

The quality of written communication will be assessed in your answer to this question.

Write your answer on separate paper or in your exercise book.

Total [6]

3 Scientists wanted to find out whether smoking increases the risk of developing an eye disease called AMD. Most AMD sufferers are partially blind.

a The scientists recorded the following statements.

Statement	
A The greater the number of years a person smokes, and the more they smoke each day, the greater their risk of developing AMD.	
B Eye doctors have noticed a build-up of waste substances near the retinas of smokers' eyes.	
C 12% of AMD sufferers smoked 20 cigarettes a day for more than 40 years.	
D Substances in cigarette smoke may cause damage to cells in the retina of the eye.	
E Eye doctors have observed that people with AMD have damaged retinas.	

Write the letter **D** next to the **three** statements that are data.

Write the letter **E** next to the **one** statement that is part of an explanation.

Write the letter **C** next to the **one** statement that is a conclusion drawn from data. [3]

b The scientists used their explanation to make the prediction shown in the box:

> Passive smokers have a greater risk of developing AMD than people who are not exposed to cigarette smoke.

They collected the following data:

- Of 100 non-smokers with AMD, 72 were passive smokers.
- Of 100 non-smokers without AMD, 66 were passive smokers.

Put a tick in the **two** boxes next to the statements that are true.

The data increases confidence in the explanation. ☐

The data proves the explanation is correct. ☐

The data agrees with the prediction. ☐

The data decreases confidence in the explanation. ☐ [2]

Total [5]

3

4 A scientist developed an explanation that doing exercise would decrease the effect that a high-salt diet has on raising blood pressure.

The scientist predicted that:

The more active you are, the less your blood pressure will increase if you switch to a high-salt diet.

The scientist collected data:

- She measured the blood pressure of 1000 volunteers who normally had a low-salt diet.
- She asked the volunteers to eat a high-salt diet for one week. She divided the volunteers into four groups. During this week, each volunteer did a different amount of exercise.
- At the end of the week, the scientist measured the volunteers' blood pressure again.

The results are in the table.

Group of volunteers	Change in blood pressure after eating a high-salt diet for one week (mmHg)
least active	5.27
next-to-least active	5.07
next-to-most active	4.93
most active	3.88

Do you think other scientists should accept or reject the explanation?

Give reasons for your decision.

____

____

____

____

Total [4]

The scientific community

1

2

3

4

The scientific community

1 Write a C next to reasons why two scientists may come to different conclusions about the same data.

Write an X next to reasons for scientists not abandoning an explanation even when new data do not seem to support the explanation.

a The scientists are interested in different areas of science. ☐

b The data may be incorrect. ☐

c The new explanation may run into problems. ☐

d Different organisations paid for each scientist's research. ☐

e It is safer to stick with ideas that have served well in the past. ☐

Exam tip

Remember – a scientist's judgements may be influenced by their:
- background
- experience
- interests
- funding

2 Read the information in the box.

> It's not only fatty foods and smoking that are risk factors for heart disease! New research shows that decaffeinated coffee may also be bad for your heart. American scientists studied 187 people for three months. They found increased levels of harmful cholesterol in the blood of people who drank decaffeinated coffee. The researchers presented their findings to other scientists at the American Heart Association's conference.

Make up a dialogue on the next page to get across six important points about the scientific community.

Use the information in the box above and the phrases below.
- scientific journal or conference
- peer review
- evaluated by other scientists
- reproducible results
- different conclusions about the same data

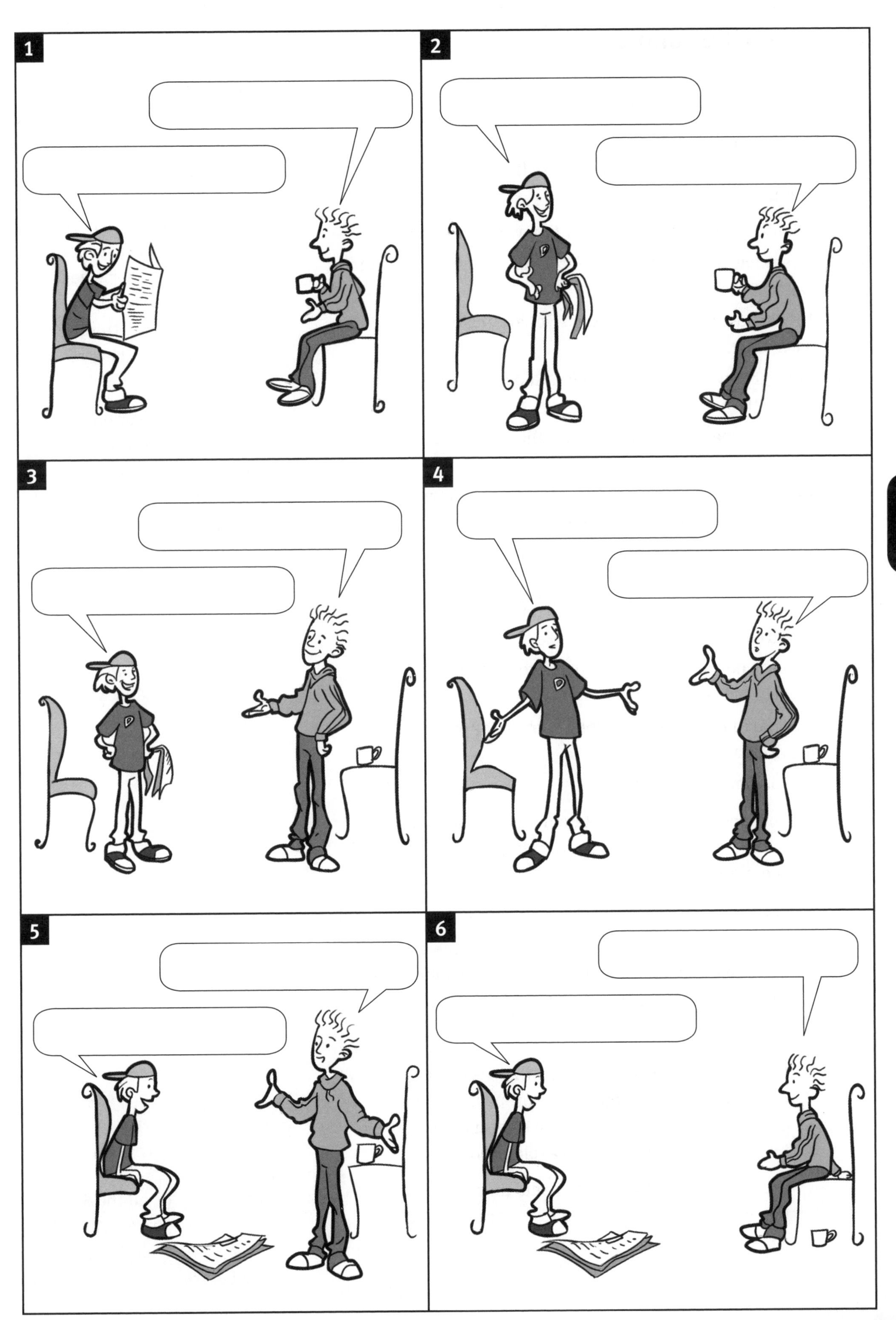
1
2
3
4
5
6

The scientific community

1 Read the information in the box opposite about the work of Charles Darwin in the 1860s.

Below are some statements from other scientists about Darwin's theory. Some statements are from 1870; others are from 2012.

A I don't know why there is variation within a species.

B There has not yet been time for other scientists to evaluate Darwin's ideas.

C DNA evidence explains how species are related to each other.

D I disagree with Darwin's theory. God created everything on Earth in seven days.

E We know that the Earth is about 4 thousand million years old.

F Darwin's observations are fine. But I don't know how living things pass on variations to their offspring.

G We often find new fossils, and we can date them accurately now.

H Darwin's theory has worked well for more than a hundred years.

I There is not enough fossil evidence to support Darwin's ideas.

J The Earth has not existed long enough for evolution to have happened.

K The data may be incorrect.

a Give the letters of four statements that show why many scientists in 1870 did not support Darwin's theory of natural selection.

______ ______ ______ ______ [4]

b Give the letters of three statements that mention **evidence** that might make a 2012 scientist more likely to accept Darwin's theory than an 1870 scientist.

______ ______ ______ [3]

c In 2030, a scientist collects data that seems to contradict Darwin's theory. He proposes a new explanation for evolution. Most scientists do not immediately accept the new explanation.

Give the letters of two statements that are reasons for scientists to not accept the new explanation.

______ ______ [2]

Total [9]

Darwin collected evidence by making these observations:

- The individuals of a species are slightly different from each other. This is variation.
- There are always more members of a species than can survive. So there is competition between members of a species.

From his observations, Darwin developed his **theory of natural selection**:

- 'Any variation that helps an individual to survive is more likely to be inherited by its offspring.'

Darwin's theory helps to explain how and why evolution happens.

Exam tip

Remember that a good explanation:

- accounts for all the data
- might explain a link that people had not thought of before.

2 Read the article below.

Alcohol and the risk of heart disease

Since the 1970s, research from many groups of scientists has suggested that drinking up to 30 g alcohol a day reduces the risk of heart disease. The researchers explained that alcohol helps prevents blood clots.

In a 2005 study, a group of scientists claimed that drinking alcohol does not reduce the risk of heart disease. The scientists concluded that the reduced risk of heart disease in moderate drinkers could be caused by other factors. For example, 27 of 30 other risk factors for heart disease are higher in non-drinkers than in moderate drinkers.

a All the scientists who researched alcohol and heart disease published reports of their findings in scientific journals.

The reports include the scientists' investigation methods, their data, and their conclusions.

i Other scientists looked carefully at the reports **before** they were published. Suggest why.

____ [1]

ii Give two reasons for publishing reports in scientific journals.

____

____ [1]

b At first, many scientists did not accept the findings of the 2005 scientists. Suggest why.

____

____ [1]

c A second group of scientists collected data that suggests the same conclusion as the 2005 research.
How might the findings of the second group influence other scientists?
Tick the **one** best answer.

Their findings make other scientists less likely to accept the claim of the 2005 scientists. ☐

Their findings prove the claim of the 2005 scientists. ☐

Their findings make other scientists more likely to question the claim of the 2005 scientists. ☐

Their findings make other scientists more likely to accept the claim of the 2005 scientists. ☐

[1]

4

d Some organisations commented on the 2005 research findings.

Draw lines to match each comment to an organisation that might have made the comment.

Organisation
Wine-making company
Organisation that persuades people not to drink alcohol
European Health Organisation

Comment
Just two studies show that moderate drinking does not reduce the risk of heart disease. Many more studies show the opposite. People can continue drinking as usual.
We need more information before we can assess whether the claim of the 2005 research is correct. We want more scientists to do research about the impact of alcohol on heart health.
The 2005 study shows that the results of the earlier research could well be wrong. People who drink alcohol are not protecting themselves against heart disease.

[3]

Total [7]

3 Between 1996 and 2001 scientists asked 1 million women to help with their health research as part of the Million Women Study.

The women wrote down their height, weight, and how much they exercised. Over the next few years, about 2500 of the women broke (fractured) their hips.

The scientists wanted to find out if there were correlations between:

- exercise and hip fracture risk
- body mass index (BMI) and hip fracture risk.

a Before starting their research, the scientists read about other scientists' research on hip fracture risks.

Put ticks next to the **two** best reasons for doing this.

Reason	Tick (✓)
to plan how to find out if the other scientists' work is reproducible	
to repeat exactly what other scientists have done	
to find as many mistakes as possible in the other scientists' investigations	
to find out what is already known about the effects of exercise and BMI on hip fracture risk	

[2]

b The Million Women Study scientists found that women who exercise are less likely to break their hips than women who do not exercise.

Other scientists made similar claims in earlier research.

Explain why the Million Women Study research increases confidence in the claims of the scientists who did the earlier research.

____

____ [1]

c The Million Women Study scientists wrote a paper about hip fracture risks for a scientific journal.

Before the paper was published, it was peer reviewed.

Outline the process of peer review.

____

____

____ [2]

d Why might the scientists tell TV and radio journalists about their work?

Put a tick next to the **one** best reason for doing this.

Reason	Tick (✓)
so that other scientists can speak about their work at a scientific conference	
to let women know that exercising can reduce their risk of hip fractures	
so that other scientists can evaluate the quality of the explanation	
so that other scientists can try to reproduce their findings	

[1]

Total [6]

4 The following is an extract from a newspaper article.

> In a ground-breaking study scientists used new techniques to measure the effect of nanoparticles of copper on the ability of stem cells to differentiate into a range of tissues.
>
> Scientists took stem cells from a specially created human embryo and cultivated the cells in a suspension of copper nanoparticles. The scientists made the following claims:
>
> **A** The cells grew and divided six times more quickly than cells cultivated in the normal way.
>
> **B** One type of cell was able to differentiate into all the tissues found in the human body.
>
> **C** The cells could be manipulated into growing into a functioning human heart.
>
> **D** All the cells, tissues, and organs produced by this technique appeared perfectly normal.

a Where do scientists report their findings?

______________________________ [1]

b Which of the statements above is definitely not based on data?

______________________________ [1]

c Suggest why a scientist reading this report might not believe all the claims.

____

______________________________ [3]

d A scientist who is sceptical might try to repeat the research carried out. Explain why this would be done.

____

______________________________ [2]

e Why might some people have ethical concerns about the work that has been carried out?

______________________________ [1]

Total [8]

Risk

Why are you lying out there in the sun at midday? Where's your sunscreen? It's dangerous to sunbathe – ultraviolet radiation from the sun hugely increases your chance of getting skin cancer. Each year, nearly 6000 British people get melanoma skin cancer. And skin cancer kills.

*Yes, but there are **benefits** too. Sunlight helps you make vitamin D. You need this to strengthen your bones and muscles, and to boost your immune system. Anyway, I feel more confident when I've got a tan. And it's lovely and warm out here...*

OK. I guess it's up to you. No-one is forcing you to take the risk. But think of the consequences of getting skin cancer. And the costs to the NHS of treating you. We'll all pay in the end.

That sandwich you are eating was labelled as GM.
Really? I didn't notice, I just love BLT sandwiches.
But GM means genetically modified – aren't you worried about all those genes in the bread or bacon that shouldn't be there?
Surely there may be some risk – I may be eating genes from a bacterium or fungus! Could I get infected?
Don't worry, all GM food is tested to be sure that the new genes cannot cause any harm.
No chance at all – all genes are made from the same simple chemicals. They all get digested.
True, but what you really should consider is what may happen if the GM animal carrying the gene breeds with wild animals and the gene is spread into the wild.
So the risk isn't to me personally, but the risk of GM is to the natural ecosystem – boy that is a big risk!
So the sandwich carries no more risk than any other sandwich.

Risk

1 Tick the box against those statements that represent risks.

- **A** A new herbicide may cause cancer. ☐
- **B** Energy is always lost as heat. ☐
- **C** Modified genes may escape to the wild population. ☐
- **D** No-one knows what effect nanoparticles may have. ☐
- **E** A new exercise may cause a muscle strain. ☐
- **F** Mineral nutrients are recycled by bacteria. ☐
- **G** Transfats used in baking are new man-made fats. ☐

2 Below are eight answers. Make up one question for each answer.

a The chance of it occurring in a large sample over a given period of time.

____

b The chance of it happening and the consequences if it did.

____

c Take account of both its risks and benefits.

____

d If they have chosen to do it.

____

e Because they are more familiar with riding a bike.

____

f Actions that may have long-term effects.

____

H

g Perceived risk.

____

h They may overestimate the risk.

____

3 Use the clues to fill in the grid.

1 Everything we do carries a risk of accident or h . . .

2 We can assess the size of a risk by measuring the c . . . of it happening in a large sample over a certain time.

3 New vaccines are an example of a scientific a . . . that brings with it new risks.

4 Radioactive materials emit i . . . radiation all the time.

5 The chance of a nuclear power station exploding is small. The c . . . of this happening would be devastating.

6 Governments or public bodies may have to a . . . what level of risk is acceptable in a particular situation.

7 Some decisions about risk may be c . . ., especially if those most at risk are not those who benefit.

8 Sometimes people think the size of a risk is bigger than it really is. Their perception of the size of the risk is greater than the s . . . calculated risk.

9 Nuclear power stations emit less carbon dioxide than coal-fired power stations. Some people think that this b . . . is worth the risk of building nuclear power stations.

10 Many people think that the size of the risk of flying in an aeroplane is greater than it really is. They p . . . that flying is risky because they don't fly very often.

11 It is impossible to reduce risk to zero. So people must decide what level of risk is a . . .

4 Draw a line to link two words on the circle.
Write a sentence on the line saying how the two words are connected.
Repeat for as many pairs as you can.

risk

safe

benefits

chance

consequences

balance

unfamiliar

scientific advances

long-lasting effects

statistically estimated risk

perceived risk

controversial

Risk

1 Read the information in the box.

Mephedrone is an illegal drug. It is similar to cocaine. It began to be widely used in 2007. It was banned in the UK in 2010. In 2011, the Government reported that about 4.4% of 14–16-year-olds had used the drug during the past year.

Mephedrone makes some users feel happy for up to 3 hours.

The drug has many side-effects. The bar charts show the findings from two surveys on the side-effects of taking mephedrone.

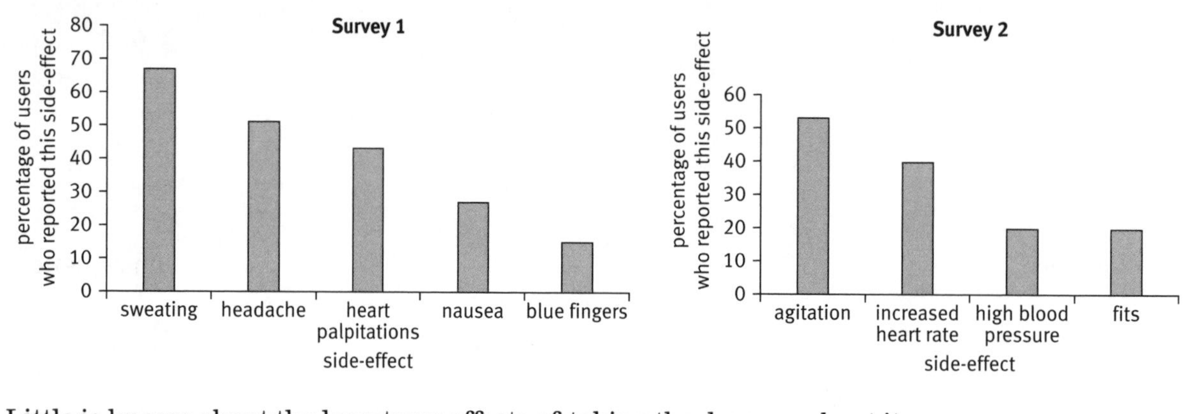

Little is known about the long-term effects of taking the drug, or about its effects when taken in combination with other drugs. One user was admitted to a psychiatric hospital after taking the drug for 18 months.

a Suggest two reasons that people may give for taking mephedrone, even though it is illegal and there are many risks associated with taking the drug.

____

____ [2]

b Kyle uses mephedrone.

He tells a radio programme, "It is my choice to use mephedrone. There are no known long-term risks. The short-term risks are ones that I am prepared to take."

Identify two reasons for Kyle's willingness to accept the risks of taking mephedrone.

____

____ [2]

c Suggest why little is known about the long-term effects of taking mephedrone.

____ [1]

d Use the bar charts to identify the three side-effects that a user of mephedrone has the highest chance of experiencing.

____ [1]

Total [6]

2 Multiple sclerosis (MS) is a disease that damages nerves of the central nervous system. People with MS may suffer pain, stiffness, trembling, and exhaustion.

Some people with MS say that smoking cannabis makes them feel better. The boxes below describe some scientific research about cannabis and MS.

Box A	**Box B**	**Box C**
Mice with MS symptoms were injected with cannabis extract. Their stiffness and trembling decreased compared to a control group.	160 people with MS were given either a cannabis extract spray or a placebo spray. Afterwards, muscle stiffness was significantly less in the cannabis spray group.	112 people with MS who smoked cannabis filled in a questionnaire. 97% of the people said that cannabis reduced their muscle stiffness.
Box D	**Box E**	**Box F**
Cannabis can produce severe anxiety and panic. The more that is smoked, the worse the anxiety and panic.	Small amounts of cannabis cause slowed reaction times, poor coordination, and short-term memory problems.	Nine pilots smoked either a joint of cannabis or a placebo. 24 hours later they did a test in a flight simulator. The cannabis pilots made more mistakes than the placebo pilots.

a Use the evidence in the boxes to discuss the benefits and risks of smoking cannabis to a person with MS.

____

____

____

____

____

______________________________ [4]

b In the UK, smoking cannabis is illegal.

Suggest why some people think that it should be legal for people with MS to smoke cannabis.

____

______________________________ [1]

c Read the opinions below about people with MS smoking cannabis.

Verity

If I drive after smoking cannabis, and then crash my car, people could be killed.

Will

I am 21. If I smoke cannabis, I am twice as likely to suffer from hallucinations and frightening thoughts than if I don't smoke cannabis.

Xena

If I smoke cannabis regularly, and then stop, I might become aggressive and hurt someone.

Yasmin

When smoking cannabis, my heart rate could double compared to normal.

i Give the name or names of the person or people who are talking about the chance of a risk occurring.

_______________________________________ [1]

ii Give the name or names of the person or people who mention the consequences of a risk.

_______________________________________ [1]

Total [7]

Making decisions: using animals for drug testing

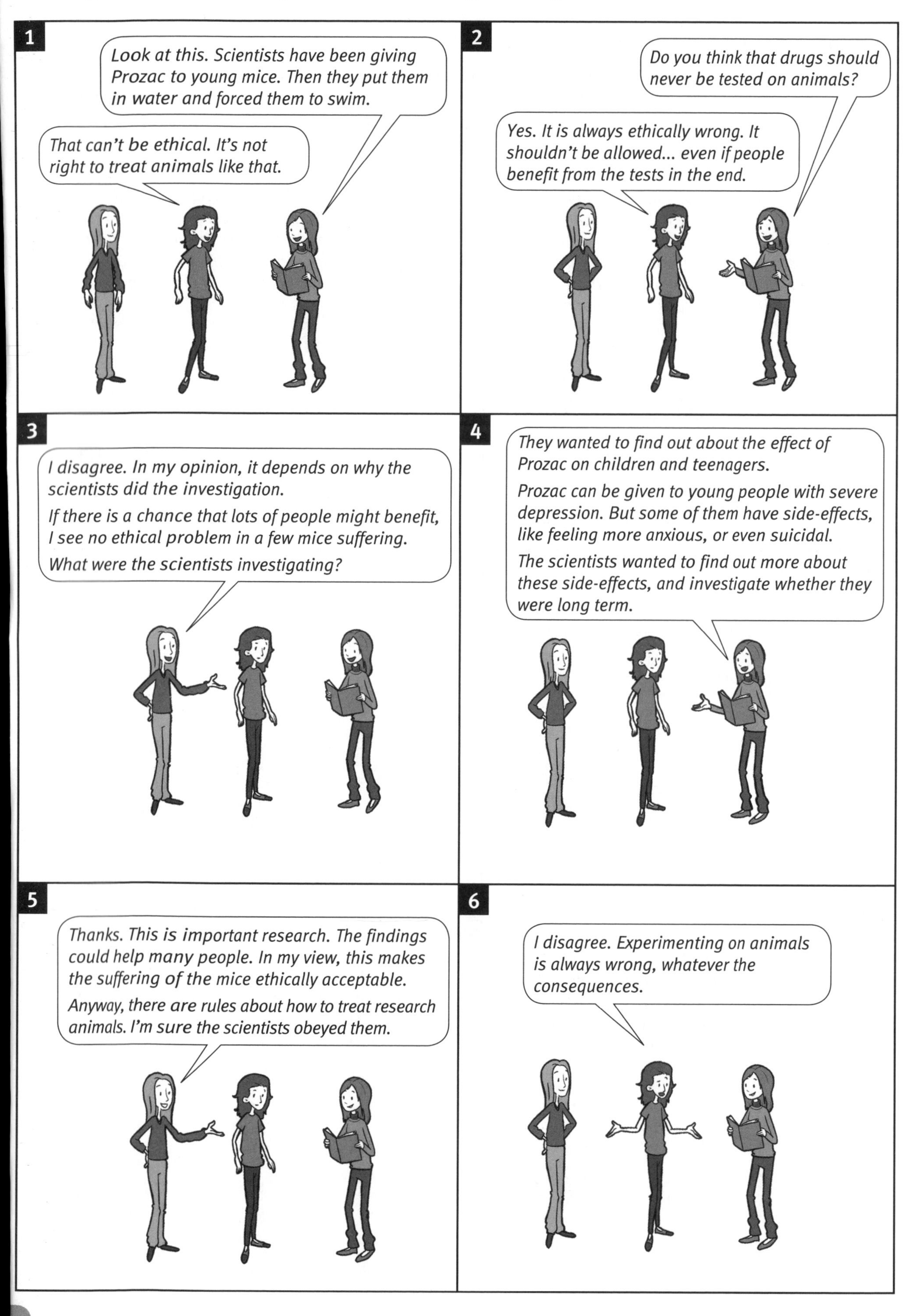

Making decisions about science and technology

Debating Society: 18 March

Is it ethically acceptable to clone human embryos to produce stem cells to treat illnesses?

Making decisions about science and technology

1 Read the information in the box about values and ethics in science.

Then make notes about values and ethics in science in the table.

- Write a title in the top row.
- Write the two or three most important points in the next row down.
- Write other, detailed, information in the lowest three rows.

There are many questions that cannot be addressed using a scientific approach. For example, a scientist can find out how to get stem cells from embryos, and how to use the stem cells to treat diseases. But people have different views about whether it is ethically right to actually use these techniques. So it is up to others – not just scientists – to answer the question *'Is it ethically acceptable to use embryonic stem cells to treat disease?'*

People use different sorts of arguments when they discuss ethical issues. One argument is that the right decision is the one that gives the best outcome for the greatest number of people involved. Another argument is that some actions are always right or wrong, whatever the consequences.

Title:	
Most important points:	
Other information:	

2 Match each statement to the idea it shows. Draw lines to match the statements to the ideas.

Statement
All 11–12-year-old girls should be vaccinated against cervical cancer.
No-one should be vaccinated against cervical cancer. It interferes with nature.
1340 out of 700 000 girls given the cervical cancer vaccine complained of side-effects. Most were minor effects such as rashes and a feeling of dizziness.
1300 girls suffer side-effects. Doctors believe 700 lives could be saved in the future.

Idea it shows
Things that are unnatural are never right.
This decision is the one that benefits the most people.
This statement compares the risk with the potential benefit.
This statement assesses the risk of having the cervical cancer vaccine.

3 Tick the boxes next to the questions that science could try to answer.

A Should Britain build more nuclear power stations? ☐

B Can nuclear power meet our energy needs? ☐

C Can nanotechnology keep food fresh? ☐

D Is it ethical to use human embryonic stem cells? ☐

E Should the UK recycle more plastic waste? ☐

F Is organic food more nutritious? ☐

G Can intensive farming grow more food? ☐

H Is it possible to develop a malaria vaccine? ☐

I If a malaria vaccine were available, who should be given it? ☐

Making decisions about science and technology

1 Pre-implantation genetic screening (PGS) is a technique to choose the best embryo to implant in a woman who is having fertility treatment.

The technique involves removing one cell from an eight-cell embryo.
Scientists then test this cell for abnormal chromosomes.
Abnormal chromosomes may cause conditions like Down's syndrome.

a Five people were asked for their opinions about PGS. Some people support the technique; others think it should not be allowed.

Write the names of the people in the correct columns in the table.

People who *agree* with pre-implantation genetic screening	People who *disagree* with pre-implantation genetic screening

[3]

b Write **S** next to the questions that could be answered using a scientific approach.

Write **V** next to the questions that address issues of **values** of PGS.

Is PGS ethically acceptable? ☐

Does the technique damage embryos? ☐

Is it natural to choose which embryo to implant in a woman's uterus? ☐

Do embryos that have been tested grow properly when they are implanted into a woman's uterus? ☐

Is PGS necessary – maybe embryos can fix their own genetic defects? ☐

Is it right to destroy embryos that are not implanted? ☐ [3]

H **c** Suggest two reasons why PGS is offered to parents in the UK but is not offered to parents in some other countries.

____

____

____

____

______________________________ [2]

Total [8]

2 Read the article in the box.

British cancer chief to investigate breast cancer screening

The Government's cancer chief has set up an inquiry to investigate the advantages and disadvantages of breast cancer screening. Currently, women over 50 are offered breast cancer screening every three years. More than 80% of breast cancer cases are diagnosed in women aged 50–69.

A review of scientific research in 2002 found that breast cancer screening in Sweden reduced the number of breast cancer deaths by 21%.

But in 2011 other scientists found that breast cancer screening leads to unnecessary treatment. Some cancers get better by themselves, for example.

Some cancer treatments, including surgery, can have bad side-effects. There is also a tiny risk that the radiation used in the screening could cause a cancer.

a From the article, identify the arguments for and against breast cancer screening.

____

____

____

____ [2]

H

b Use the data in the table, and information in the box at the start of this question, to answer the question below.

	Life expectancy (years)	Estimated population with HIV/AIDS (%)
Swaziland	32	26
UK	79	0.14

Suggest why the government of Swaziland might decide to spend more money diagnosing HIV/AIDs than on breast cancer screening.

____

____

____ [2]

Total [4]

3 All British babies are offered the MMR vaccine against measles, mumps, and rubella.

Some parents decide not to have their babies vaccinated.

a Give one reason why some people think that parents should be free to choose whether or not their baby gets the MMR vaccine.

____

____

____ [1]

b Give one reason why some people think that the MMR vaccine should be compulsory.

____

____

____ [1]

Total [2]

4 Organic farming is a sustainable means of producing food. Discuss the reasons why farmers in the UK might move to organic farming, while farmers in India might be advised not to change to organic farming.

✎ The quality of written communication will be assessed in your answer.

Total [6]

accuracy An accurate instrument or procedure gives a 'true' reading.

active site The part of an enzyme that the reacting molecules fit into.

H **active transport** Molecules are moved into or out of a cell using energy. This process is used when transport needs to be faster than diffusion, and when molecules are being moved from a region where they are at low concentration to where they are at high concentration.

adaptation Features that help an organism survive in its environment.

ADH A hormone making kidney tubules more permeable to water, causing greater re-absorption of water.

aerobic respiration Respiration that uses oxygen.

alcohol The intoxicating chemical in wine, beer, and spirits. Causes changes in behaviour and may create long-term addiction.

algae Simple green water plants.

algal bloom Rapid growth of algae, making the water green. It can be toxic.

allele Different versions of the same gene.

Alzheimer's disease A form of senile dementia caused by irreversible degeneration of the brain.

amino acids The small molecules that are joined in long chains to make proteins. All the proteins in living things are made from 20 different amino acids joined in a different order.

anaerobic Without oxygen.

anaerobic respiration Respiration that does not use oxygen.

H **antagonistic effectors** Antagonistic effectors have opposite effects.

antagonistic pair Two muscles that work to move the same bone in opposite directions, for example, the biceps and triceps muscles.

antibiotic Drug that kills or stops the growth of bacteria and fungi.

antibiotic resistant Microorganisms that are not killed by antibiotics.

antibodies A group of proteins made by white blood cells to fight dangerous microorganisms. A different antibody is needed to fight each different type of microorganism. Antibodies bind to the surface of the microorganism, which triggers other white blood cells to digest them.

antigens Proteins on the surface of a cell. A cell's antigens are unique markers.

aorta The main artery that carries oxygenated blood away from the left ventricle of the heart.

aquaculture Farming in water, such as fish farming.

arrhythmia A problem with the heart in which the muscle does not contract regularly – the rhythm is lost.

arteries Blood vessels that carry blood away from the heart.

asexual reproduction When a new individual is produced from just one parent.

assumption A piece of information that is taken for granted without sufficient evidence to be certain.

atrium (plural atria) One of the upper chambers in the heart. The two atria pump blood to the ventricles.

H **auxin** A plant hormone that affects plant growth and development. For example, auxin stimulates growth of roots in cuttings.

axon A long, thin extension of the cytoplasm of a neuron. The axon carries electrical impulses very quickly.

bacteria Single-celled microorganisms that do not have a nucleus. Some bacteria may cause disease.

bacteriophage A type of virus that infects bacteria.

base pairing The bases in a DNA molecule (A, C, G, T) always bond in the same way. A and T always bond together. C and G always bond together.

baseline data Data gathered at the start of a study or experiment so that patterns and trends can be established.

behaviour Everything an organism does; its response to all the stimuli around it.

best estimate When measuring a variable, the best estimate is the value in which you have most confidence.

beta blockers Drugs that block the receptor sites for the hormone adrenaline. They inhibit the normal effects of adrenaline on the body.

H **bioaccumulation** Build-up of chemicals in organisms as the chemicals travel through the food chain.

biodegradable Substances that can be broken down by microorganisms such as bacteria and fungi. Most paper and wood items are biodegradable, but most plastics are not.

biodiversity The great variety of living things, both within a species and between different species.

bioethanol Ethanol fuel produced by yeast fermentation of plant materials, such as cane sugar and sugar beet.

biofuels Fuel made from crops such as rape seed.

biogas Methane gas produced by the anaerobic digestion of organic material, such as farm-animal manure. Biogas contains a significant proportion of carbon dioxide.

blind trial A clinical trial in which the patient does not know whether they are taking the new drug, but their doctor does.

blood pressure The pressure exerted by blood pushing on the walls of a blood vessel.

body mass index Your body mass index is calculated using the formula BMI = body mass (kg) / $[\text{height (m)}]^2$. Tables will tell you if your body mass is healthy for your size.

bone Strong, rigid tissues making up the skeleton of vertebrates.

H **calculated risk** Risk calculated from reliable data.

cancer A growth or tumour caused by abnormal and uncontrolled cell division.

capillary Tiny blood vessel that is one cell thick. Capillaries carry blood through the tissues between the arteries and veins.

H **capillary bed** Large numbers of narrow blood vessels that pass through each organ in the body. Capillaries receive blood from arteries and return it to veins. Capillary walls are only one cell thick.

carbohydrate A natural chemical made of carbon, hydrogen, and oxygen. An example is glucose, $C_6H_{12}O_6$. Carbohydrates include sugars, starch, and cellulose.

carbon cycle The cycling of the element carbon in the environment between the atmosphere, biosphere, hydrosphere, and lithosphere. The element exists in different compounds in these spheres. In the atmosphere it is mainly present as carbon dioxide.

carbon sink A system taking carbon dioxide from the air and storing it, for example, a growing forest.

carrier Someone who has the recessive allele for a characteristic or disease but who does not have the characteristic or disease itself.

cartilage Tough, flexible tissue found at the end of bones and in joints. It protects the end of bones from rubbing together and becoming damaged.

catalyst Chemical that starts or speeds up the rate of a reaction without being changed by it.

cause When there is evidence that changes in a factor produce a particular outcome, then the factor is said to cause the outcome, for example, increases in the pollen count cause increases in the incidence of hayfever.

cell The basic structural and functional unit of all living things.

cell membrane Thin layer surrounding the cytoplasm of a cell. It restricts the passage of substances into and out of the cell.

cell wall Rigid outer layer of plant cells and bacteria.

cellulose The chemical that makes up most of the fibre in food. The human body cannot digest cellulose.

central nervous system In mammals this is the brain and spinal cord.

cerebral cortex The highly folded outer region of the brain, concerned with conscious behaviour.

chlorophyll A green pigment found in chloroplasts. Chlorophyll absorbs energy from sunlight for photosynthesis.

chloroplast An organelle found in some plant cells where photosynthesis takes place.

chromosome Long, thin, threadlike structures in the nucleus of a cell made from a molecule of DNA. Chromosomes carry the genes.

chymosin Enzyme found in calves' stomachs that breaks down proteins (a protease). Fungi have been genetically modified to produce chymosin industrially for cheese-making.

clinical trial When a new drug is tested on humans to find out whether it is safe and whether it works.

clone A new cell or individual made by asexual reproduction. A clone has the same genes as its parent.

closed-loop system A system with no waste – everything is recycled.

cloud formation Evaporation of water, for example, from a forest, condensing into clouds.

combustion The process of burning a substance that reacts with oxygen to produce heat and light.

competition Different organisms that require the same resource, such as water, food, light, or space, must compete for the resource.

concentrated solution The concentration of a solution depends on how much dissolved chemical (solute) there is compared with the solvent. A concentrated solution contains a high level of solute to solvent.

conditioned reflex A reflex where the response is associated with a secondary stimulus, for example, a dog salivates when it hears a bell because it has associated the bell with food.

conditioning Reinforcement of behaviour associated with conditioned reflexes.

conscious To have awareness of surroundings and sensations.

consciousness The part of the human brain concerned with thought and decision making.

consumers Organisms that eat others in a food chain. This is all the organisms in a food chain except the producer(s).

control In a clinical trial, the control group is people taking the currently used drug. The effects of the new drug can then be compared to this group.

core (of the body) Central parts of the body where the body temperature is kept constant.

coronary artery Artery that supplies blood carrying oxygen and glucose directly to the muscle cells of the heart.

coronary heart disease A disease where the coronary arteries become increasingly blocked with fatty deposits, restricting the blood flow to the heart muscle. The risk of this is increased by a high-fat diet, smoking, and drinking excess alcohol.

correlation A link between two things, for example, if an outcome happens when a factor is present, but not when it is absent, or if an outcome increases or decreases when a factor increases. For example, when pollen count increases hayfever cases also increase.

crop rotation Changing the crop grown in a field each year to preserve fertility.

crude oil Oil straight from an oil well, not refined into petrol or diesel.

cutting A shoot or leaf taken from a plant, to be grown into a new plant.

cystic fibrosis An inherited disorder. The disorder is caused by recessive alleles.

cytoplasm Gel enclosed by the cell membrane that contains the cell organelles such as mitochondria.

dead organic matter Any material that was once part of a living organism.

decomposer Organism that feeds on dead organisms. Decomposers break down the complex organic chemicals in their bodies, releasing nutrients back into the ecosystem to be used by other living organisms.

deforestation Cutting down and clearing forests leaving bare ground.

dehydration Drying out.

denatured A change in the usual nature of something. When enzymes are denatured by heat, their structure, including the shape of the active site, is altered.

H **denitrification** Removal of nitrogen from soil. Bacteria break down nitrates in the soil, converting them back to nitrogen.

H **denitrifying bacteria** Bacteria that break down nitrates in the soil, releasing nitrogen into the air.

deoxygenated Blood in which the haemoglobin is not bound to oxygen molecules.

desert Very dry area where no plants can grow. The area can be cold or hot.

desertification Turning to desert.

H **detritivore** Organism that feeds on dead organisms and waste. Woodlice, earthworms, and millipedes are examples of detritivores.

development How an organism changes as it grows and matures. As a zygote develops, it forms more and more cells. These are organised into different tissues and organs.

diabetes type 1 An illness where the level of sugar in the blood cannot be controlled. Type 1 diabetes starts suddenly, often when people are young. Cells in the pancreas stop producing insulin. Treatment is by regular insulin injections.

diabetes type 2 An illness where the level of sugar in the blood cannot be controlled. Type 2 diabetes develops in people with poor diets or who are obese. The cells in the body stop responding to insulin. Treatment is through careful diet and regular exercise.

diastolic The blood pressure when all parts of the heart muscle are relaxed and the heart is filling with blood.

differentiated A differentiated cell has a specialised form suited to its function. It cannot change into another kind of cell.

diffusion Movement of molecules from a region of high concentration to a region of lower concentration.

digest Break down larger, insoluble molecules into small, soluble molecules.

digestive enzyme Biological catalysts that break down food.

dilute The concentration of a solution depends on how much dissolved chemical (solute) there is compared with the solvent. A dilute solution contains a low level of solute to solvent.

dioxin Poisonous chemicals, for example, released when plastics burn.

direct drilling Planting seeds directly into the soil without ploughing first.

disease A condition that impairs normal functioning of an organism's body, usually associated with particular signs and symptoms. It may be caused by an infection or by the dysfunction of internal organs.

dislocation An injury where a bone is forced out of its joint.

DNA (deoxyribonucleic acid) The chemical that makes up chromosomes. DNA carries the genetic code, which controls how an organism develops.

DNA fingerprinting A DNA fingerprint uses gene probes to identify particular sequences of DNA bases in a person's genetic make-up. The pattern produced in a DNA fingerprint can be used to identify family relationships.

DNA profiling A DNA profile is produced in the same way as a DNA fingerprint, but fewer gene probes are used. DNA profiling is used in forensic science to test samples of DNA left at crime scenes.

dominant Describes an allele that will show up in an organism even if a different allele of the gene is present. You only need to have one copy of a dominant allele to have the feature it produces.

double circulation A circulatory system where the blood passes through the heart twice for every complete circulation of the body.

double helix The shape of the DNA molecule, with two strands twisted together in a spiral.

H **double-blind trial** A clinical trial in which neither the doctor nor the patient knows whether the patient is taking the new drug.

H **economic context** How money changes hands between businesses, government, and individuals.

ecosystem Living organisms plus their non-living environment working together.

ecosystem services Life-support systems that we depend on for our survival.

Ecstasy A recreational drug that increases the concentration of serotonin at the synapses in the brain, giving pleasurable feelings. Long-term effects may include destruction of the synapses.

effector The part of a control system that brings about a change to the system.

embryo The earliest stage of development for an animal or plant. In humans the embryo stage lasts for the first two months.

embryo selection A process whereby an embryo's genes are checked before the embryo is put into the mother's womb. Only healthy embryos are chosen.

embryonic stem cell Unspecialised cell in the very early embryo that can divide to form any type of cell, or even a whole new individual. In human embryos the cells are identical and unspecialised up to the eight-cell stage.

endangered Species that are at risk of becoming extinct.

environment Everything that surrounds you. This includes factors such as the air and water, as well as other living things.

enzyme A protein that catalyses (speeds up) chemical reactions in living things.

epidemiological study Scientific study that examines the causes, spread, and control of a disease in a human population.

ethanol Waste product of anaerobic respiration in plants and yeast.

ethical Non-scientific, concerned with what is right or wrong.

ethics A set of principles that may show how to behave in a situation.

H **eutrophication** Build-up of nutrients in water.

evolution The process by which species gradually change over time. Evolution can produce new species.

excretion The removal of waste products of chemical reactions from cells.

extinct A species is extinct when all the members of the species have died out.

factor A variable that changes and may affect something else.

fallow crop Crop that is not harvested, allowing the field to regain nutrients.

false negative A wrong test result. The test result says that a person does not have a medical condition but this is incorrect.

false positive A wrong test result. The test result says that a person has a medical condition but this is incorrect.

fatty sheath Fat wrapped around the outside of an axon to insulate neurons from each other.

feral Untamed, wild.

fermentation Chemical reactions in living organisms that release energy from organic chemicals, such as yeast producing alcohol from the sugar in grapes.

fermenter Large vessel in which microorganisms are grown to make a useful product.

fertile An organism that can produce offspring.

fitness State of health and strength of the body.

flowers Reproductive structures in plants often containing both male and female reproductive structures.

fluorescent marker Chemical attached to a DNA strand so it can be found or identified when separated from other strands in a gel.

food chain In the food industry this covers all the stages from where food grows, through harvesting, processing, preservation, and cooking to being eaten.

food web A series of linked food chains showing the feeding relationships in a habitat – 'what eats what'.

fossil The stony remains of an animal or plant that lived millions of years ago, or an imprint of its mark, for example, a footprint, in a surface.

fossil fuel Fuel made of the bodies of long-dead organisms.

fossil sunlight energy Sunlight energy stored as chemical energy in fossil fuel.

fruit Remaining parts of a flower containing seeds after fertilisation.

fungus (plural fungi) A group of living things, including some microorganisms, that cannot make their own food.

gametes The sex cells that fuse to form a zygote. In humans, the male gamete is the sperm and the female gamete is the egg.

gas exchange The exchange of oxygen and carbon dioxide that takes place in the lungs.

gene A section of DNA giving the instructions for a cell about how to make one kind of protein.

gene probe A short piece of single-stranded DNA used in a genetic test. The gene probe has complementary bases to the allele that is being tested for.

gene switching Genes in the nucleus of a cell switch off and are inactive when a cell becomes specialised. Only genes that the cell needs to carry out its particular job stay active.

genetic Factors that are affected by an organism's genes.

genetic modification (GM) Altering the characteristics of an organism by introducing the genes of another organism into its DNA.

genetic screening Testing a population for a particular allele.

genetic study Scientific study of the genes carried by people in a population to look for alleles that increase the risk of disease.

genetic variation Differences between individuals caused by differences in their genes. Gametes show genetic variation – they all have different genes.

genotype A description of the genes an organism has.

glands Parts of the body that make enzymes, hormones, and other secretions in the body, for example, sweat glands.

glucose Sugar produced during photosynthesis.

habitat The place where an organism lives.

haemoglobin The protein molecule in red blood cells. Haemoglobin binds to oxygen and carries it around the body. It also gives blood its red colour.

heart disease A range of potentially serious illnesses that affect the heart.

heavy metals Metals such as lead and mercury, which are toxic in small concentrations.

herbicide Chemical that kills plants, usually plants that are weeds in crops or gardens.

herbicide resistant Plants that are not killed by herbicides.

heterozygous An individual with two different alleles for a particular gene.

homozygous An individual whose alleles for a particular gene are the same.

hormone A chemical messenger secreted by specialised cells in animals and plants. Hormones bring about changes in cells or tissues in different parts of the animal or plant.

human trial The stage of the trial process for a new drug where the drug is taken by healthy volunteers to see if it is safe, and then by sick volunteers to check that it works.

Huntington's disease An inherited disease of the nervous system. The symptoms do not show up until middle age.

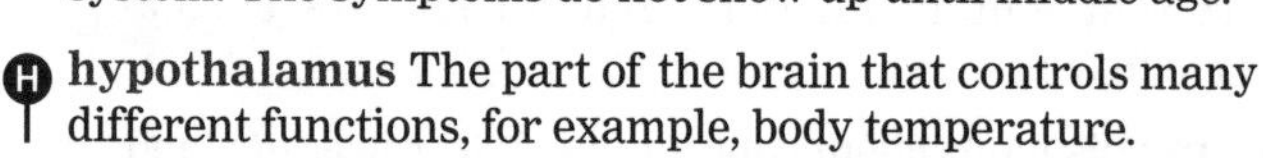

hypothalamus The part of the brain that controls many different functions, for example, body temperature.

immune Able to react to an infection quickly, stopping the microorganisms before they can make you ill, usually because you've been exposed to them before.

immune system A group of organs and tissues in the body that fight infections.

infectious A disease that can be caught. The microorganism that causes it is passed from one person to another through the air, through water, or by touch.

infertile An organism that cannot produce offspring.

inherited A feature that is passed from parents to offspring by their genes.

innate Inborn, inherited from parents via genes.

insoluble Does not form a solution (dissolve) in water or other solutes.

insulin A hormone produced by the pancreas. It is a chemical that helps to control the level of sugar (glucose) in the blood.

intensive agriculture Farming with high inputs of fertiliser and pesticides and high productivity.

involuntary An automatic response made by the body without conscious thought.

joint A point where two or more bones meet.

kidney Organ in the body that removes waste urea from the blood, and balances water and blood plasma levels. People are usually born with two kidneys.

lactic acid Waste product from anaerobic respiration in animals.

learn To gain new knowledge or skills.

lichen Organism consisting of a fungus growing with a simple photosynthetic organism called an alga. Lichens grow very slowly and are often found growing on walls and roofs.

life cycle The stages an organism goes through as it matures, develops, and reproduces.

lifestyle The way in which people choose to live their lives, for example, what they choose to eat, how much exercise they choose to do, how much stress they experience in their job.

lifestyle disease Disease that is not caused by microorganisms. It is triggered by other factors, for example, smoking, diet, and lack of exercise.

lifestyle history The way you have been living, taking regular exercise, eating healthily, and so on.

ligament Tissue that joins two or more bones together.

light intensity The amount of light reaching a given area.

light meter Device for measuring light intensity.

lignocellulase Enzyme that can break down the woody fibres in plant material (lignin) and the cellulose of plant cell walls.

limiting factor The factor that prevents the rate of photosynthesis from increasing at a particular time. This may be light intensity, temperature, carbon dioxide concentration, or water availability.

linear system A system based on the take–make–dump model.

lock-and-key model In chemical reactions catalysed by enzymes, molecules taking part in the reaction fit exactly into the enzyme's active site. The active site will not fit other molecules – it is specific. This is like a key fitting into a lock.

long-term memory The part of the memory that stores information for a long period, or permanently.

match Some studies into diseases compare two groups of people. People in each group are chosen to be as similar as possible (matched) so that the results can be fairly compared.

mayfly larvae Mayflies spend most of their lives (up to three years) as larvae (also called mayfly nymphs). They live and feed in aquatic environments. The adult insects live on the wing for a short time, from a few hours to a few days.

medical history Health or health problems in the past.

medication Any pharmaceutical drug used to treat or prevent an illness.

meiosis Cell division that halves the number of chromosomes to produce gametes. The four new cells are genetically different from each other and from the parent cell.

memory The storage and retrieval of information by the brain.

memory cell Long-lived white blood cell, which is able to respond very quickly (by producing antibodies to destroy the microorganism) when it meets a microorganism for the second time.

meristem cells Unspecialised cells in plants that can develop into any kind of specialised cell.

microorganism Living organism that can only be seen by looking through a microscope. They include bacteria, viruses, and fungi.

mitochondrion (plural mitochondria) An organelle in animal and plant cells where respiration takes place.

mitosis Cell division that makes two new cells that are genetically identical to each other and to the parent cell.

models of memory Explanations for how memory is structured in the brain.

monoculture The continuous growing of one type of crop.

motor neuron A neuron that carries nerve impulses from the brain or spinal cord to an effector.

H **mRNA** Messenger RNA, a chemical involved in making proteins in cells. The mRNA molecule is similar to DNA but single stranded. It carries the genetic code from the DNA molecule out of the nucleus into the cytoplasm.

multistore model One explanation for how the human memory works.

muscles Muscles move parts of the skeleton for movement. There is also muscle tissue in other parts of the body, for example, in the walls of arteries.

mutation A change in the DNA of an organism. It alters a gene and may change the organism's characteristics.

nanometre A unit of measurement (abbreviation nm). A millimetre is the same as 1 million nanometres. $1\ \text{nm} = 1 \times 10^{-3}\ \mu\text{m} = 1 \times 10^{-6}\ \text{mm} = 1 \times 10^{-9}\ \text{m}$.

native species Organisms naturally occurring in an area – not introduced by humans.

natural selection When certain individuals are better suited to their environment they are more likely to survive and breed, passing on their features to the next generation.

H **negative feedback** A system where any change results in actions that reverse the original change.

nerve cell A cell in the nervous system that transmits electrical signals to allow communication within the body.

nerve impulses Electrical signals carried by neurons (nerve cells).

nervous system Tissues and organs that control the body's responses to stimuli. In a mammal it is made up of the central nervous system and peripheral nervous system.

neuron Nerve cell.

neuroscientist A scientist who studies how the brain and nerves function.

newborn reflexes Reflexes to particular stimuli that usually occur only for a short time in newborn babies.

nitrate ions An ion is an electrically charged atom or group of atoms. The nitrate ion has a negative charge, NO_3^-.

nitrogen cycle The continual cycling of nitrogen, which is one of the elements essential for life. By being converted to different chemical forms, nitrogen is able to pass between the atmosphere, lithosphere, hydrosphere, and biosphere.

H **nitrogen fixation** When nitrogen in the air is converted into nitrates in the soil by bacteria.

nitrogen-fixing bacteria Bacteria found in the soil and in swellings (nodules) on the roots of some plants (legumes), such as clover and peas. These bacteria take in nitrogen gas and make nitrates, which plants can absorb and use to make proteins.

non-biodegradable Waste materials that microorganisms cannot break down.

nucleus Central structure in a cell containing genetic material. It controls the function and characteristics of the cell.

obesity A medical condition where the increase in body fat poses a serious threat to health. A body mass index over 30 kg/m^2.

H **open-label trial** A clinical drug test in which both the patient and their doctor know whether the patient is taking the new drug.

optimum temperature The temperature at which enzymes work fastest.

organ Part of a plant or animal made up of different tissues.

organelles The specialised parts of a cell, such as the nucleus and mitochondria. Chloroplasts are organelles that occur only in plant cells.

osmosis The diffusion of water across a partially permeable membrane.

outcome A variable that changes as a result of something else changing.

outlier A measured result that seems very different from other repeat measurements, or from the value you would expect, which you therefore strongly suspect is wrong.

overgrazing Too many grazing animals, such as goats, damaging the environment.

oxygenated Blood in which the haemoglobin is bound to oxygen molecules (oxyhaemoglobin).

pancreas An organ in the body that produces some hormones and digestive enzymes. The hormone insulin is made here.

partially permeable membrane A membrane that acts as a barrier to some molecules but allows others to diffuse through freely.

pathway A series of connected neurones that allow nerve impulses to travel along a particular route very quickly.

H **perceived risk** The level of risk that people think is attached to an activity, not based on data.

peripheral nervous system The network of nerves connecting the central nervous system to the rest of the body.

phagocytosis Engulfing and digestion of microorganisms and other foreign matter by white blood cells.

H **phenotype** A description of the physical characteristics that an organism has (often related to a particular gene).

phloem A plant tissue that transports sugar throughout a plant.

photosynthesis The process in green plants that uses energy from sunlight to convert carbon dioxide and water into the sugar glucose.

phototropism The bending of growing plant shoots towards the light.

phytoplankton Single-celled photosynthetic organisms found in an ocean ecosystem.

H **pituitary gland** Part of the human brain that coordinates many different functions, for example, release of ADH.

plasma The clear straw-coloured fluid part of blood.

plasmids Small circle of DNA found in bacteria. Plasmids are not part of a bacterium's main chromosome.

platelets Cell fragments found in blood. Platelets play a role in the clotting process.

pollen Plant reproductive structures containing a male gamete.

pollinators Animals, such as bees, that transfer pollen from anther to stigma.

polymer A material made up of very long molecules. The molecules are long chains of smaller molecules.

population A group of animals or plants of the same species living in the same area.

predator An animal that kills other animals (its prey) for food.

H **pre-implantation genetic diagnosis (PGD)** This is the technical term for embryo selection. Embryos fertilised outside the body are tested for genetic disorders. Only healthy embryos are put into the mother's uterus.

primary forest A forest that has never been felled or logged.

processing centre The part of a control system that receives and processes information from the receptor, and triggers action by the effectors.

producers Organisms found at the start of a food chain. Producers are autotrophs, able to make their own food.

proportional Two variables are proportional if there is a constant ratio between them.

protein Chemicals in living things that are polymers made by joining together amino acids.

Prozac A brand name for an anti-depressant drug. It increases the concentration of serotonin at the synapses in the brain.

pulmonary artery The artery that carries deoxygenated blood to the lungs. The artery leaves the right ventricle of the heart.

pulmonary vein The vein that carries oxygenated blood from the lungs to the left atrium of the heart.

pulse rate The rate at which the heart beats. The pulse is measured by pressing on an artery in the neck, wrist, or groin.

pupil reflex The reaction of the muscles in the pupil to light. The pupil contracts in bright light and relaxes in dim light.

quadrat A square grid of a known area that is used to survey plants in a location. Quadrats come in different sizes up to 1 m^2. The size of quadrat that is chosen depends on the size of the plants and also the area that needs to be surveyed.

quota Agreed total amount that can be taken or harvested per year.

random Of no predictable pattern.

range The difference between the highest and the lowest of a set of data.

rate of photosynthesis Rate at which green plants convert carbon dioxide and water to glucose in the presence of light.

reactants Substances used in reactions by living organisms or by non-living matter.

H **real difference** One way of deciding if there is a real difference between two values is to look at the mean values and the ranges. The difference between two mean values is real if their ranges do not overlap.

receptor The part of a control system that detects changes in the system and passes this information to the processing centre.

receptor molecule A protein (often embedded in a cell membrane) that exactly fits with a specific molecule, bringing about a reaction in the cell.

recessive An allele that will only show up in an organism when a dominant allele of the gene is not present. You must have two copies of a recessive allele to have the feature it produces.

recovery period The time for you to recover after taking exercise and for your heart rate to return to its resting rate.

red blood cells Blood cells containing haemoglobin, which binds to oxygen so that it can be carried around the body by the bloodstream.

reflex arc A neuron pathway that brings about a reflex response. A reflex arc involves a sensory neuron, connecting neurons in the brain or spinal cord, and a motor neuron.

reject How a body might react to foreign material introduced in a transplant.

relay neuron A neuron that carries the impulses from the sensory neuron to the motor neuron.

rennet An enzyme used in cheese-making.

repeatable A quality of a measurement that gives the same result when repeated under the same conditions.

H **reproducible** A quality of a measurement that gives the same results when carried out under different conditions, for example, by different people or using different equipment or methods.

repetition Act of repeating.

repetition of information Saying or writing the same thing several times.

reproductive isolation Two populations are reproductively isolated if they are unable to breed with each other.

respiration A series of chemical reactions in cells that release energy for the cell to use.

response Action or behaviour that is caused by a stimulus.

retina Light-sensitive layer at the back of the eye. The retina detects light by converting light into nerve impulses.

retrieval of information Collecting information from a particular source.

ribosomes Organelles in cells. Amino acids are joined together to form proteins in the ribosomes.

RICE RICE stands for rest, ice, compression, elevation. This is the treatment for a sprain.

risk A measure of the size of a potential danger. It is calculated by combining a measure of a hazard with the chance of it happening.

risk factor A variable linked to an increased risk of disease. Risk factors are linked to disease but may not be the cause of the disease.

root hair cell Microscopic cell that increases the surface area for absorption of minerals and water by plant roots.

rooting powder A product used in gardening containing plant hormones. Rooting powder encourages a cutting to form roots.

sample Small part of something that is likely to represent the whole.

selective breeding Choosing parent organisms with certain characteristics and mating them to try to produce offspring that have these characteristics.

sensory neuron A neuron that carries nerve impulses from a receptor to the brain or spinal cord.

serotonin A chemical released at one type of synapse in the brain, resulting in feelings of pleasure.

sex cell Cells produced by males and females for reproduction – sperm cells and egg cells. Sex cells carry a copy of the parent's genetic information. They join together at fertilisation.

sexual reproduction Reproduction where the sex cells from two individuals fuse together to form an embryo.

shivering Very quick muscle contractions. Releases more energy from muscle cells to raise body temperature.

short-term memory The part of the memory that stores information for a short time.

silting of rivers Eroded soil making the water muddy and settling on the river bed.

simple reflex An automatic response made by an animal to a stimulus.

single-celled protein (SCP) A microorganism grown as a source of food protein. Most single-celled protein is used in animal feed, but one type is used in food for humans.

skeleton The bones that form a framework for the body. The skeleton supports and protects the internal organs, and provides a system of levers that allow the body to move. Some bones also make red blood cells.

social behaviour Behaviour that takes place between members of the same species, including humans.

social context The situation of people's lives.

soil erosion Soil removal by wind or rain into rivers or the sea.

specialised A specialised cell is adapted for a particular job.

species A group or organisms that can breed to produce fertile offspring.

sprain An injury where ligaments are located.

stable ecosystem An ecosystem that renews itself and does not change.

starch A type of carbohydrate found in bread, potatoes, and rice. Plants produce starch to store the energy food they make by photosynthesis. Starch molecules are a long chain of glucose molecules.

starch grains Microscopic granules of starch forming an energy store in plant cells.

stem cell Unspecialised animal cell that can divide and develop into specialised cells.

stimulus A change in the environment that causes a response.

stomata Tiny holes in the underside of a leaf that allow carbon dioxide into the leaf and water vapour and oxygen out of the leaf.

structural Making up the structure (of a cell or organism).

structural proteins Proteins that are used to build cells.

sustainability Using resources and the environment to meet the needs of people today without damaging Earth or reducing the resources for people in the future.

sustainable Able to continue over long periods of time.

symptom What a person has when they have a particular illness, for example, a rash, high temperature, or sore throat.

synapse A tiny gap between neurons that transmits nerve impulses from one neuron to another by means of chemicals diffusing across the gap.

synovial fluid Fluid found in the cavity of a joint. The fluid lubricates and nourishes the joint, and prevents two bones from rubbing against each other.

tendon Tissue that joins muscle to a bone.

termination When medicine or surgical treatment is used to end a pregnancy.

theory A scientific explanation that is generally accepted by the scientific community.

H **therapeutic cloning** Growing new tissues and organs from cloning embryonic stem cells. The new tissues and organs are used to treat people who are ill or injured.

tissue Group of specialised cells of the same type working together to do a particular job.

tissue fluid Plasma that is forced out of the blood as it passes through a capillary network. Tissue fluid carries dissolved chemicals from the blood to cells.

torn ligament An injury of the elastic tissues that hold bones together, a common sports injury of the knee. For treatment see 'RICE'.

torn tendon An injury of the inelastic tissues that connect muscles to bones. For treatment see 'RICE'.

toxic Poisonous.

transect A straight line that runs through a location. Data on plant and animal distribution is recorded at regular intervals along the line.

transmitter substance Chemical that bridges the gap between two neurons.

triplet code A sequence of three bases coding for a particular amino acid in the genetic code.

uncertain Describes measurements where scientists know that they may not have recorded the true value.

uncertainty The amount by which a measurement could differ from the true value.

unspecialised Cells that have not yet developed into one particular type of cell.

vaccination Introducing to the body a chemical (a vaccine) used to make a person immune to a disease. A vaccine contains weakened or dead microorganisms, or parts of the microorganism, so that the body makes antibodies to the disease without being ill.

valves Flaps of tissue that act like one-way gates, only letting blood flow in one direction around the body. Valves are found in the heart and in veins.

variation Differences between living organisms. This could be differences between species. There are also differences between members of a population from the same species.

H **vasoconstriction** Narrowing of blood vessels.

H **vasodilation** Widening of blood vessels.

vector A method of transfer. Vectors are used to transfer genes from one organism to another.

vein Blood vessel that carries blood towards the heart.

vena cava The main vein that returns deoxygenated blood to the right atrium of the heart.

ventricle One of the lower chambers of the heart. The right ventricle pumps blood to the lungs. The left ventricle pumps blood to the rest of the body.

virus Microorganisms that can only live and reproduce inside living cells.

white blood cells Cells in the blood that fight microorganisms. Some white blood cells digest invading microorganisms. Others produce antibodies.

working memory The system in the brain responsible for holding and manipulating information needed to carry out tasks.

XX chromosomes The pair of sex chromosomes found in a human female's body cells.

XY chromosomes The pair of sex chromosomes found in a human male's body cells.

xylem Plant tissue that transports water through a plant.

yeast Single-celled fungus used in brewing and baking.

yield The crop yield is the amount of crop that can be grown per area of land.

zygote The cell made when a sperm cell fertilises an egg cell in sexual reproduction.

B1 Workout

1 **a** Cell **b** Genes **c** Nucleus **d** Chromosome
e DNA

2 Dimples – one gene only
Weight – genes and environment
Eye colour – several genes working together
Scars – environment only

3 **a** T **b** T **c** F
d F **e** F

4 Unspecialised, asexual, clones, environments

5 **a** If the test is positive, should I have an abortion?
b If the test is positive, should I father a child?

6 **a**

	R	r
R	RR	Rr
r	Rr	rr

b **i** 75%
ii 25%

7 –

B1 Quickfire

1 Chromosomes, genes, instructions, proteins, DNA

2 True statements: **a, b**
Corrected versions of false statements:
c If a person has one dominant allele in a pair of alleles, they will show the characteristic linked to that gene.
d Human male body cells have XY sex chromosomes.
e A sperm contains chromosomes that are not paired up.

3 Thick mucus, difficulty breathing, chest infections, digestion problems

4 Is the father a carrier of cystic fibrosis? Would she consider having a termination if the fetus had cystic fibrosis? Is she prepared for the increased risk of miscarriage when cells are removed from the fetus for testing?

5 **a** Ellie, Susan and Tom
b

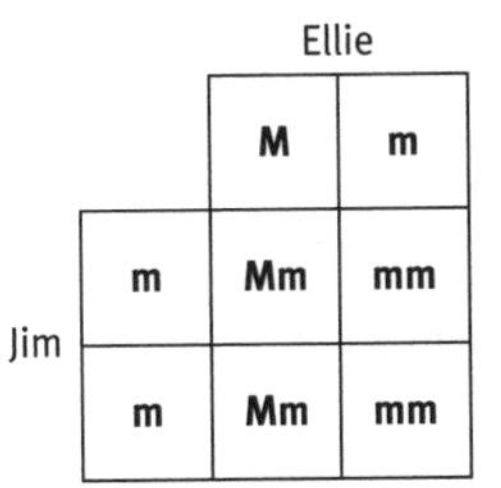

		Ellie	
		M	m
Jim	m	Mm	mm
	m	Mm	mm

The chance their baby has Marfan syndrome is 50%.

6 **a** **Rr** **b** 2, 6, 7, 10 **c** 3, 4, 5, 8, 9, 11

7 Tremor, clumsiness, memory loss, inability to concentrate, mood changes

8 A stem cell is an unspecialised cell. Embryonic stem cells are extracted from embryos. They can develop into any type of cell. Adult stem cells can develop into many, but not all, types of cells.

9 **a** A person's genotype describes a person's genes. So the genotype of people represented by those on the top line is XX. The phenotype describes the characteristics of a living organism. The phenotype of those on the top line is that they are female.
b Two offspring genotypes are XX (phenotype female) and two genotypes are XY (phenotype male). So there is a 50% chance of a baby being female.

10 **a** She has dimples.
b Dominant – she has dimples even though she has only one **D** allele.
c Heterozygous – she has two different versions of the dimple allele.

11 The Y chromosome includes a sex-determining gene. This makes an embryo develop testes, and so become male. When there is no Y chromosome, the embryo develops ovaries. It is female.

12 **a** Pre-implantation genetic diagnosis involves creating embryos outside the body, and testing them for an inherited disorder. An embryo that will not develop the inherited disorder is selected for implantation into the uterus
b If a man and a woman are carriers of a genetic disease
c Whether they would be prepared to abort a fetus with the genetic disease; whether removing a cell to test damages an embryo

13 Scientists remove an egg cell nucleus. They take a nucleus from an adult body cell of the organism they want to clone, and transfer it to the 'empty' egg cell. They grow the embryo for a few days and implant it into a uterus.

B1 GCSE-style questions

1 **a** **i** They have the same combination of alleles; they both developed from one egg that was fertilised by one sperm;
ii Their cells certain 23 pairs of chromosomes.
iii XX
b **i** They have different lifestyles.
ii A stem cell is an unspecialised cell.
iii To treat diseases

2 **a** **i** 50%
ii 2
b **i**

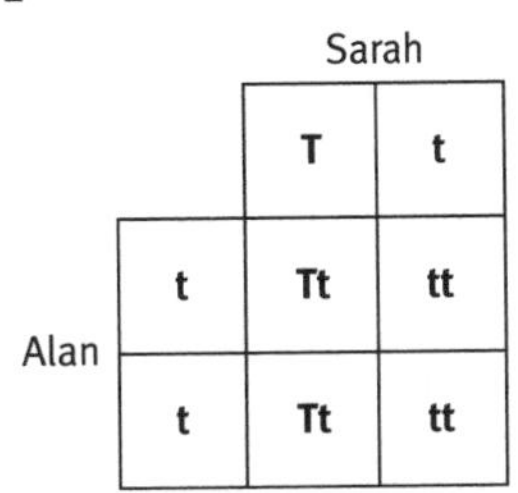

		Sarah	
		T	t
Alan	t	Tt	tt
	t	Tt	tt

ii 50%

3 **a** Two from: tremor, clumsiness, memory loss, inability to concentrate, mood changes
b Abigail and Brenda
c For the test: they can consider having a termination if the test is positive. Against the test: deciding whether or not to have a termination is a very difficult decision.

4 5/6 marks
Answer clearly explains why siblings show variation **and** explains why children have different characteristics from their parents.
All information in the answer is relevant, clear, organised and presented in a structured and coherent format. Specialist terms are used appropriately. Few, if any, errors in grammar, punctuation and spelling.
3/4 marks
Answer explains clearly **either** why siblings show variation **or** why children have different characteristics from their parents.
OR explains each point above but lacks detail/clarity.
Most of the information is relevant and presented in a structured and coherent format. Specialist terms are usually used correctly. There are occasional errors in grammar, punctuation and spelling.
1/2 marks
Answer explains **either** why siblings show variation **or** why children have different characteristics from their parents.
AND the explanation lacks detail/clarity.
There may be limited use of specialist terms. Errors of grammar, punctuation and spelling prevent communication

of the science. Answer includes 1 or 2 points of those listed below.
0 marks
Insufficient or irrelevant science. Answer not worthy of credit.
Relevant points include:

- Children inherit alleles from both parents
- There are two possible types of allele for each gene
- A child inherits alleles when a sex cell (sperm) from his or her dad fertilizes a sex cell (egg) from his or her mum
- The sex cells contained one version of each allele from each parent – this explains why children have different characteristics from their parents
- Alleles pair up on fertilization, so body cells contain pairs of alleles
- The combinations of alleles in each pair determine the characteristics of a child
- Different children in a family (except identical twins) inherit different combinations of alleles from their parents – this is why siblings show variation

5 a A child can get the disease even if their parents did not have the disease.
b Rod, Barbara, Sally, Philip
c Mitch and Tracy
d Phenotype describes the observable characteristics of the person, including unsteady walking, slurring of speech, and reduced appetite. Genotype describes the combination of alleles an organism has. A person with Niemann-Pick disorder has two copies of the faulty gene. The alleles from both parents are identical.
e i Pre-implantation genetic diagnosis involves removing one cell from each of several embryos made outside the body. The cells are then tested for faulty alleles. An embryo without faulty alleles is chosen for implantation in the uterus.
ii Implications include: the possibility that removing a cell to test damages the embryo; the ethical question of whether it is right to destroy embryos with faulty alleles; the possibility of tests providing false positive or false negative results.

B2 Workout

1 Top empty box: reproduce rapidly
Empty boxes on middle line: make toxins and damage cells
Bottom empty box: disease symptoms
2 a Experts meet every April...
b The eggs provide food...
c Technicians break...
d This flu virus is delivered...
3 2C, 3D, 4K, 5E, 6F, 8G, 9H, 10L, 11I, 12J

B2 Quickfire

1 **b**, **c**
2 a G b L c L
d L e L f G
3 Pulse rate – number of heart beats per minute
Lower blood pressure measurement – pressure against artery wall when heart is relaxed
Higher blood pressure measurement – pressure against artery wall when heart is contracting
4 True statements: **b**, **c**, **e**
Corrected versions of false statements:
a Antibiotics kill bacteria.
d In clinical trials, new drugs are tested for effectiveness on people with the illness.

5

Part of circulation system	What does it do?	What is it made from?
heart	pumps blood around the body	muscle
artery	takes blood from your heart to the rest of your body	thick walls made from muscle and elastic fibres
vein	brings blood back to your heart from the rest of your body	thin walls made of muscle and elastic fibres
capillary	takes blood to and from tissues, and allows oxygen and food to diffuse into cells and waste to diffuse out of cells	very thin walls (one cell thick)

6 A, C, F, D, E, G, B
7 Receptor – detects changes in the environment
Processing centre – receives information and processes responses
Effector – produces the response
8 Vaccines and medicines may have unwanted side-effects.
9 Complete the whole course of treatment; only take antibiotics when necessary.
10 a Neither P nor D b P c D, P
11 Long-term human trials for new drugs ensure that the drugs are safe, and that they work. The trials may also identify side-effects that do not occur immediately.
12 Antimicrobial resistance develops when random changes in bacteria or fungi genes make new varieties that the antimicrobial cannot kill or inhibit.
13 Temperature sensor – receptors in brain; thermostat system with switch – pituitary gland; heater – ADH and kidneys
14 Alcohol suppresses ADH, which leads to the production of a greater volume of more dilute urine. Ecstasy leads to increased ADH production, which results in a smaller volume of more concentrated urine.

B2 GCSE-style questions

1 a i Reproduction
ii 4
b i White
ii Taking anti-diarrhoea tablets would mean that *Salmonella* bacteria would stay in the intestines for longer.
2 a Increased, opinion B *or* C *or* D, decreased, opinion A
b A, E, G, C, F, B, D
3 5/6 marks
Answer clearly explains why **AND** how water levels in the cells of a human body are kept constant
and explains how alcohol and Ecstasy affect the amount and concentration of urine.
All information in the answer is relevant, clear, organised and presented in a structured and coherent format. Specialist terms are used appropriately. Few, if any, errors in grammar, punctuation and spelling.
3/4 marks
Answer clearly explains why **AND** how water levels in the cells of a human body are kept constant
OR explains how alcohol and Ecstasy affect the amount and concentration of urine.
OR explains each point above but lacks detail/clarity.
Most of the information is relevant and presented in a

structured and coherent format. Specialist terms are usually used correctly. There are occasional errors in grammar, punctuation and spelling.
1/2 marks
Answer clearly explains why **OR** how water levels in the cells of a human body are kept constant
OR explains how alcohol and Ecstasy affect the amount and concentration of urine.
AND the explanation lacks detail/clarity.
There may be limited use of specialist terms. Errors of grammar, punctuation and spelling prevent communication of the science. Answer includes 1 or 2 points of those listed below.
0 marks
Insufficient or irrelevant science. Answer not worthy of credit.
Relevant points include:

- Cells only work properly if the concentrations of their contents are correct. So their water levels must be kept constant.
- Kidneys control water levels in the body by responding to changes in water levels in blood plasma.
- Water levels in blood plasma may decrease because of sweating, eating salty food, not drinking much water.
- When water levels in blood plasma are low, kidneys make smaller quantities of concentrated urine.
- When water levels in blood plasma are high, kidneys make bigger quantities of dilute urine.
- Alcohol leads to big volumes of dilute urine.
- Ecstasy leads to small volumes of concentrated urine.

4 5/6 marks
Answer clearly explains how a negative feedback system keeps water levels constant in the cells of a human body **and** includes the names of the hormones and organs involved in the system
All information in the answer is relevant, clear, organised and presented in a structured and coherent format. Specialist terms are used appropriately. Few, if any, errors in grammar, punctuation and spelling.
3/4 marks
Answer explains how a negative feedback system keeps water levels constant in the cells of a human body
OR states the names of the hormones and organs involved in the system
OR explains each point above but lacks detail/clarity.
Most of the information is relevant and presented in a structured and coherent format. Specialist terms are usually used correctly. There are occasional errors in grammar, punctuation and spelling.
1/2 marks
Answer explains how a negative feedback system keeps water levels constant in the cells of a human body
OR states the names of the hormones and organs involved in the system
AND the explanation lacks detail/clarity.
There may be limited use of specialist terms. Errors of grammar, punctuation and spelling prevent communication of the science. Answer includes 1 or 2 points of those listed below.
0 marks
Insufficient or irrelevant science. Answer not worthy of credit.
Relevant points include:

- Receptors in the brain detect changes in concentration of blood plasma.
- If the concentration is too high, the pituitary gland releases ADH – a hormone – into the blood stream
- The ADH travels to the kidneys – (the effectors.)
- The more ADH that arrives, the more water the kidneys reabsorb in the body..
- ...and the more concentrated the urine.
- If smaller amounts of ADH arrive at the kidneys, large quantities of dilute urine are made.

B3 Workout

1 –
2 –
3 **a** Stores: atmosphere, fossil fuels; processes: photosynthesis, combustion, respiration, eating, dying, decomposing
b

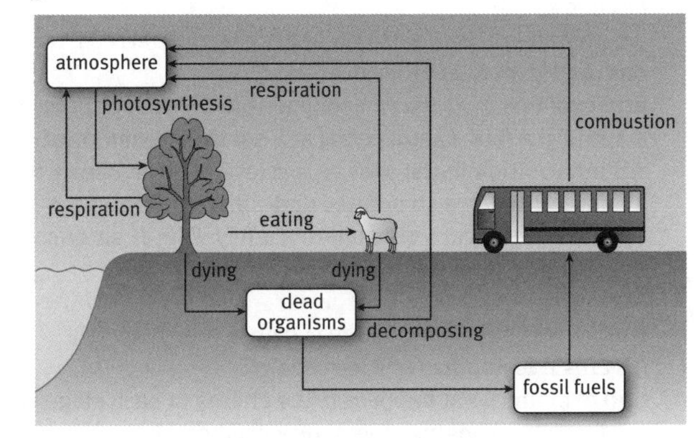

4 Animal Kingdom, Animal Kingdom, species, similar, characteristics

B3 Quickfire

1 **a** N **b** L **c** N **d** L
2 Variation – differences between organisms
Mutations – changes to genes
Habitat – the place where an organism lives
Environment – everything around an organism, including air, water, and other living things
3 Small, photosynthesis, chemicals, energy, moving, warm, heat, waste, dead, materials
4 True statements: **c**, **d**
Corrected versions of false statements:
a Life on Earth began about 3500 million years ago.
b Mutated genes in sex cells can be passed to offspring.
5 Thick fur provides insulation to help keep them warm; white colour is camouflage so their prey cannot see them easily; good swimmers mean they can move from place to place.
6 **a** Termite, kangaroo, wombat
b Grass, eucalyptus tree
c Honeyeater bird, termite, kangaroo, wombat
d The wombat population might increase since there would be less competition for its food (grass).
e The honeyeater bird population would decrease.
f The wombat population might increase since dingoes are its predator.
7 14%
8 There are changes in the environment to which a species cannot adapt; a new species arrives that competes with, eats, or causes disease of the species; another species in its food web becomes extinct.
9 Darwin's theory fits in with advances in understanding of genetics; there is no evidence or mechanism for Lamarck's ideas about the inheritance of acquired characteristics.
10 Interdependence means that species in a habitat rely on each other for food and other needs.

11 A detritivore is an organism that feeds on dead organisms and waste.

12 Decay – breaks down proteins
Nitrogen fixation – forms nitrates from nitrogen
Excretion – removes waste from an organism
Denitrification – breaks down nitrates to form nitrogen

B3 GCSE-style questions

1 a i Antarctic whelk *or* brittlestar *or Trematomus bernacchii* fish
ii Their populations will decrease.
b Environmental conditions change; a new species that is a predator of the brittlestar is introduced to Antarctica.

2 5/6 marks
Answer clearly explains how energy is transferred from one organism to another in a food chain
and explains in detail what happens to the energy at each stage of the food chain, giving at least **three** examples.
All information in the answer is relevant, clear, organised and presented in a structured and coherent format. Specialist terms are used appropriately. Few, if any, errors in grammar, punctuation and spelling.
3/4 marks
Answer explains how energy is transferred from one organism to another in a food chain
and explains what happens to the energy at each stage of the food chain, giving at least **one** example.
Most of the information is relevant and presented in a structured and coherent format. Specialist terms are usually used correctly. There are occasional errors in grammar, punctuation and spelling.
1/2 marks
Answer explains how energy is transferred from one organism to another in a food chain
or explains what happens to the energy at each stage of the food chain, giving at least **one** example.
AND the explanation lacks detail/clarity.
There may be limited use of specialist terms. Errors of grammar, punctuation and spelling prevent communication of the science. Answer includes 1 or 2 points of those listed below.
0 marks
Insufficient or irrelevant science. Answer not worthy of credit.
Relevant points include:
- Energy is transferred when animals eat other organisms.
- Energy is transferred when decay organisms eat dead organisms and waste materials.
- A small percentage of energy is passed on because the rest of the energy
  - is used for life processes (eg moving)
  - escapes to the surroundings as heat
  - is excreted and passed on to decomposers
  - cannot be eaten and is passed on to decomposers

3 a Studying fossils; analysing the DNA of modern cats, lions, and other species of the cat family.
b i Selection
ii C, B, D, A
c i 6.7 million years ago
ii Fishing cat
iii Environmental, isolation, natural, selection

B4 Workout

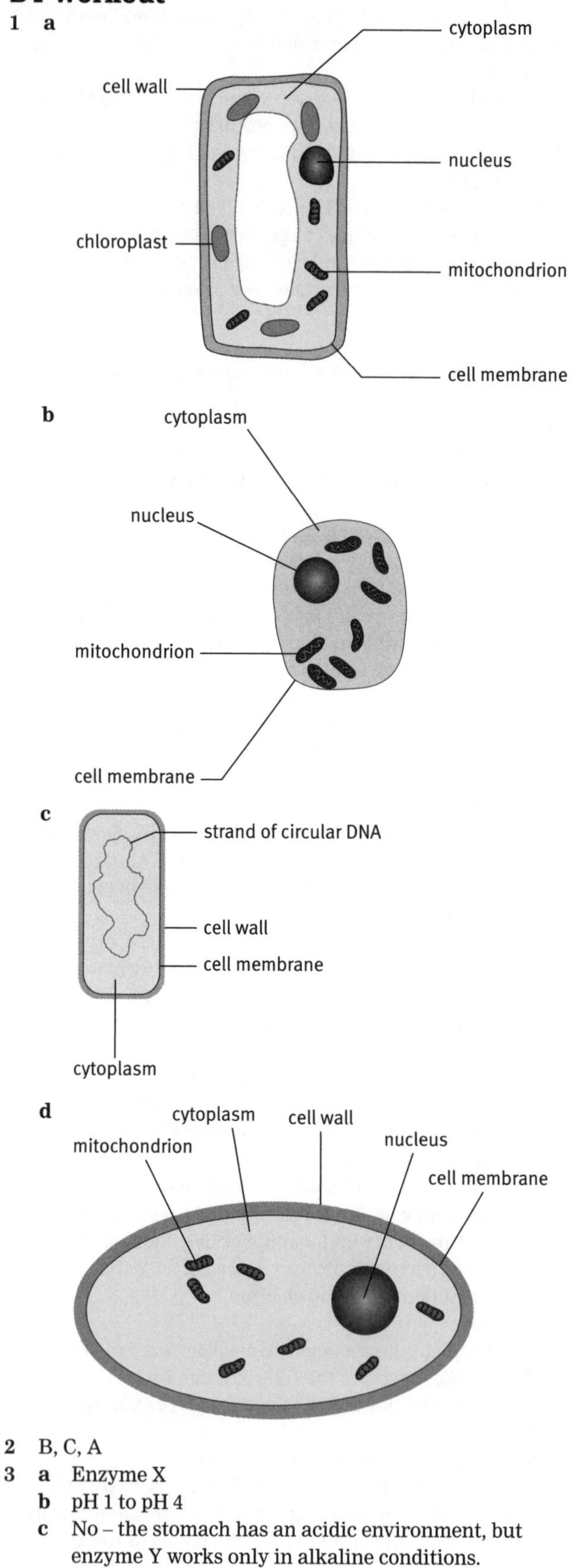

2 B, C, A

3 a Enzyme X
b pH 1 to pH 4
c No – the stomach has an acidic environment, but enzyme Y works only in alkaline conditions.

4 a A
b A
c A

d B
e A and B
f A and B

B4 Quickfire

1 Movement, respiration, sensitivity, growth, reproduction, excretion, nutrition
2 True statements: **b**, **e**
Corrected versions of false statements:
a The waste process of photosynthesis is oxygen.
c Aerobic respiration releases more energy per glucose molecule than anaerobic respiration.
d Enzymes are proteins that speed up chemical reactions.
3 a Carbon dioxide + water ⟶ glucose + oxygen
b Glucose + oxygen ⟶ carbon dioxide + water
c Glucose ⟶ lactic acid
d Glucose ⟶ ethanol + carbon dioxide
4 Diffusion: movement of molecules from a region of their higher concentration to one of their lower concentration
Osmosis: movement of water through a semi-permeable membrane to a region of their higher concentration to one of their lower concentration
Synthesis: making a chemical with bigger particles from ones with smaller particles
5 Starch, cellulose, glucose, plant, nitrates, proteins
6 In plant roots, when the soil is waterlogged; in bacteria, in deep puncture wounds; in human muscle cells during vigorous exercise
7 a 1.5 units
b 37 °C
c 10 °C to 50 °C
d The shape of the active site changes (the enzyme becomes denatured).
8 a 00:00 (midnight)
b 12:00 (midday)
c 12:00 (midday)
9 a $6CO_2 + 6H_2O \longrightarrow C_6H_{12}O_6 + 6O_2$
b $C_6H_{12}O_6 + 6O_2 \longrightarrow 6CO_2 + 6H_2O$
10 Nitrates, roots, energy, respiration

B4 GCSE-style questions

1 a As time goes on, the height of the dough increases. There is a rapid initial increase that slows to a stop at 16 cm after 20 minutes.
b i Zita's idea is partly correct – up to the optimum temperature of the enzyme, the higher the temperature, the faster the dough will rise. But above this temperature the enzyme will be denatured, so the dough will not rise at all.
ii Sam's idea is not correct. The anaerobic respiration reaction of the yeast is not affected by light intensity, though the lamp may increase the temperature.
iii Tom could add an alkali to the yeast and sugar mixture, and repeat Riana's procedure. He would need to wear eye protection and protective gloves. If his dough rises more than Riana's in the same time, then his idea would be correct.
c Glucose ⟶ ethanol + carbon dioxide
d Nucleus – the genetic code for making enzymes used in respiration is found here; cytoplasm – the enzymes used in anaerobic respiration are found here; mitochondria – the enzymes found in aerobic respiration are found here.

2 5/6 marks: answer clearly describes how substances move into and out of plant cells by diffusion, including osmosis, **and** explains the processes logically and coherently.
All information in the answer is relevant, clear, organised, and presented in a structured and coherent format. Specialist terms are used appropriately. There are few, if any, errors in grammar, punctuation, and spelling.
3/4 marks: answer describes how substances move into and out of plant cells by diffusion, including osmosis, **and** there is some logic and coherence in the answer.
Most of the information is relevant and presented in a structured and coherent format. Specialist terms are usually used correctly. There are occasional errors in grammar, punctuation, and spelling.
1/2 marks: answer briefly describes how substances move into and out of plant cells by diffusion, **but** does not make clear that osmosis is a special case of diffusion **and** the explanation lacks logic and coherence. There may be limited use of specialist terms. Errors of grammar, punctuation, and spelling prevent communication of the science. Answer includes 1 or 2 points of those listed below.
0 marks: insufficient or irrelevant science. Answer not worthy of credit.
Relevant points include:
- Diffusion is the overall movement of molecules from a region of their higher concentration to a region of their lower concentration.
- Diffusion is a passive process – it does not need extra energy.
- Carbon dioxide molecules get into leaves by diffusion.
- Oxygen molecules get out of leaves by diffusion.
- Osmosis is a special case of diffusion.
- Osmosis is the movement of water molecules from a region of their higher concentration to a region of their lower concentration through a partially permeable membrane.
- Water gets into root cells by osmosis.

3 a i Range = 15 to 18 bubbles
ii

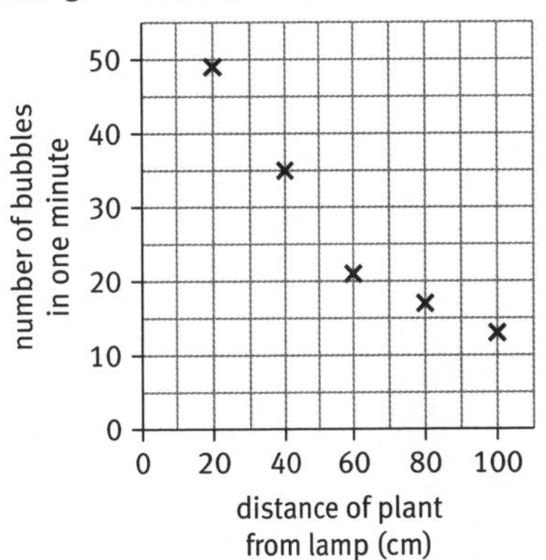

b The closer the lamp is to the plant, the greater the number of bubbles that are produced in one minute. This indicates that the rate of photosynthesis increases as the lamp gets closer.
c Artem could heat water in kettle, or in a beaker over a Bunsen burner flame, or use a thermostatically controlled water bath. He could measure the rate of bubble production by placing the pondweed in water of different temperatures. He would need a thermometer to measure the temperature of the water. He would need to keep other conditions, such as light intensity, constant.

d i Carbon dioxide + water ⟶ glucose + oxygen
ii Chloroplast

4 a Martha. To find out how plant species change gradually from one area to another
b Quadrat A because it includes foxgloves and wild garlic plants, which grow well in shade.

5 a $6CO_2 + 6H_2O \rightarrow C_6H_{12}O_6 + 6O_2$
b At 0.04% carbon dioxide, as light intensity increases, so does the rate of photosynthesis, up to a certain limit. Above this value, another factor must be limiting the rate of photosynthesis. At the higher concentration of carbon dioxide, the reaction is faster. At this higher concentration, rate increases as light intensity increases, as before. But again the graph becomes horizontal when another factor is in short supply.
c i Nitrates are used to make amino acids. These are joined together in particular orders to make proteins with different structures and functions.
ii Plant root hair cells absorb nitrates by active transport across the cell membrane. The process requires energy. First, nitrate ions enter a carrier protein in the membrane. Energy is transferred and the protein changes its shape. This releases the nitrate ion on the other side of the membrane. The carrier protein then goes back to its original shape.

B5 Workout

1 1 testes, meiosis, 4, sperm
2 23
3 ovaries, meiosis, eggs
4 23
5 fertilisation
6 zygote, 23, 46
7 mitosis, 2, 46
8 2, 4
9 4, 8
10 stem
11 16

2 Auxin, chromosomes, embryonic, fetus, gametes, mitochondria, nucleus, organelles, phototropism, tissues, unspecialised, xylem, young, zygote

3

	Meiosis	Mitosis
What does it make?	gametes (sex cells)	body cells
How many new cells does each parent cell make?	4	2
How many chromosomes are in each cell?	half as many as in the parent cell	same as in the parent cell
Where does it happen?	in sex organ cells	in body cells
Why does it happen?	to make sex cells for sexual reproduction	so an organism can grow, and replace damaged cells

B5 Quickfire

1 a P
b B
c P
d B
e B
f P

2 Organelles, mitochondria, chromosomes, DNA, strands, mitosis, chromosomes, cell

3

Species	Number of chromosomes in sex cell	Number of chromosomes in body cell
horse	32	64
wolf	39	78
carp	52	104
human (other answers are possible)	23	46

4 A – T, T – A, C – G, G – C

5 True statements: **b**, **e**, **f**, **i**, **j**
Corrected versions of false statements:
a Up to the 8-cell stage, all the cells in a human embryo are identical.
c Adult stem cells can become many, but not all, types of cell needed by a human.
d A zygote divides by mitosis to form an embryo.
g Groups of specialised cells in an animal are called tissues.
h Cell division by meiosis produces new cells that are not identical to the parent cell.

6

Cell	Are these genes switched on?		
	gene to make keratin	gene to make salivary amylase	gene for respiration
nail	yes	no	yes
hair	yes	no	yes
embryonic stem cell	yes	yes	yes
salivary gland cell	no	yes	yes
muscle cell	no	no	yes

7 a Nucleus
b Cytoplasm
c Mitochondria

8 a Light from the right; highest concentration of auxins on left of bent part of shoot
b Light from the left; highest concentration of auxins on right of bent part of shoot
c Light from above; concentration of auxins similar in all parts of shoot

9 Messenger RNA

10 The order of bases in a gene is the code for building up amino acids in the correct order to make a particular protein.

B5 GCSE-style questions

1 a Gametes, testes, 4, different, 15
b i Egg or ovum
ii Fertilisation
c C, A, D, B

2 a i They can grow many new plants quickly and cheaply; they can reproduce a plant with exactly the features they want.
ii Auxins
b i Top row: 6; bottom row: 2
ii Taking several readings means that a mean can be calculated. This is a better estimate of the quantity than taking one reading alone.
iii The cuttings with rooting powder grew more roots than those without. This suggests that rooting

powder may contain a substance that helps plant stems to grow roots.

c The graph shows that, for roots and stems, growth increases as auxin concentration increases up to a maximum concentration. Above this concentration, growth decreases steadily from the maximum. The auxin concentrations at which this happens is greater for stems than for roots.

3 a i Stem cells are unspecialised cells. They divide and develop into specialised cells.
ii They will grow heart muscle cells from the stem cells.
iii A coherent argument for or against removing and storing umbilical cord blood.
b i One of: embryos, adults
ii One of (embryos) ethical objections, problem of rejection; (adults) difficult to separate from other cells, problem of rejection

4 5/6 marks: answer describes several differences between meiosis **and** mitosis and the descriptions of differences are clear and detailed.
All information in the answer is relevant, clear, organised, and presented in a structured and coherent format. Specialist terms are used appropriately. There are few, if any, errors in grammar, punctuation, and spelling.
3/4 marks: answer gives several differences between mitosis and meiosis **but** the answer lacks detail **or** answer gives one or two differences between mitosis and meiosis **and** the descriptions of the differences are clear and detailed.
Most of the information is relevant and presented in a structured and coherent format. Specialist terms are usually used correctly. There are occasional errors in grammar, punctuation, and spelling.
1/2 marks: answer gives one or two differences between mitosis and meiosis **and** the descriptions of the differences lack detail.
There may be limited use of specialist terms. Errors of grammar, punctuation, and spelling prevent communication of the science. Answer includes 1 or 2 points of those listed below.
Relevant points include:
- Meiosis happens in sex organs; mitosis happens in all other body cells.
- Meiosis makes four new cells from one parent cell; mitosis makes two new cells.
- Meiosis makes gametes (sex cells); mitosis makes body cells.
- Meiosis makes cells with different genetic information. Mitosis makes genetically identical cells.
- Meiosis makes cells with half the number of chromosomes as are in the parent cell. Mitosis makes cells with the same number of chromosomes as the parent cell.
- Meiosis happens so that an organism can reproduce. Mitosis happens so that an organism can grow and replace damaged tissues.

5 5/6 marks: answer explains in detail why many genes are inactive in a specialised body cell **and** the explanation is logical and coherent.
All information in the answer is relevant, clear, organised, and presented in a structured and coherent format. Specialist terms are used appropriately. There are few, if any, errors in grammar, punctuation, and spelling.
3/4 marks: answer explains why many genes are inactive in a specialised body cell **but** the answer lacks detail **or** the explanation lacks logic and coherence.
Most of the information is relevant and presented in a structured and coherent format. Specialist terms are usually used correctly. There are occasional errors in grammar, punctuation, and spelling.
1/2 marks: answer briefly explains why many genes are inactive in a specialised body cell **and** the answer lacks detail **and** the explanation lacks logic and coherence. There may be limited use of specialist terms. Errors of grammar, punctuation, and spelling prevent communication of the science. Answer includes 1 or 2 points of those listed below.
0 marks: insufficient or irrelevant science. Answer not worthy of credit.
Relevant points include:
- In a specialised body cell, many genes are switched off because the cell only makes the proteins it needs for its functions.
- When cloning a mammal, inactive genes can be reactivated (switched on).
- This may happen when the nucleus of an egg cell is removed.
- The nucleus is replaced with one from the individual to be cloned.
- The egg cell divides by mitosis to makes an embryo.
- Stem cells from the embryo can be put in a dish of nutrients.
- These stem cells can develop into the type of tissue required.

B6 Workout

1 Behaviour – anything an animal does; stimulus – a change in the environment; response – an action caused by a change in the environment

2 Left from top: motor neuron, sensory neuron, effector, receptor
Right from top: spinal cord

3 Nucleus: controls cell; cytoplasm: cell reactions happen here; cell membrane: substances get into and out of the cell through this; fatty sheath: insulates neuron from neighbouring cells; branched endings: make connections with other neurons or effectors

4 Correct order: middle, top, bottom.
Notes for middle diagram: a nerve impulse gets to the end of the sensory neuron; notes for top diagram: the sensory neuron releases a chemical into the synapse and the chemical diffuses across the synapse; notes for bottom diagram: the chemical arrives at receptor molecules on the motor neuron's membrane and the chemical's molecules bind to the receptor molecules. This stimulates a nerve impulse in the motor neuron.

5 Bird: those caterpillars are poisonous. She won't eat them again because they taste so bad. Caterpillar: she's learnt that caterpillars like us don't taste good – so that's one less bird that's going to try to eat me.

B6 Quickfire

1 a S
b C
c S
d C

2 a Brain and spinal cord
b Motor neuron and sensory neuron

3 Electrical impulse – travels quickly and brings about short-term changes; hormone – travels in the blood and brings about long-term changes

4 A, F, C, D, E, B

5 a S
b S
c E
d E
e S
f S
g E

6 a Brains
b Neurons
c Consciousness, language, intelligence, memory
d Studying patients with brain damage, electrically stimulating parts of the brain, doing MRI scans

7 A, E, C, D, B, F, G

8 True statements: **b**, **d**, **f**
Corrected versions of false statements:
a Long-term memory is a seemingly limitless store of information *or* short-term memory has a limit to the information it can store.
c Repetition moves information from your short-term memory to your long-term memory.
e Your sensory memory stores memories linked to sounds and visual information.

9 a Reflex responses are automatic and very quick since no processing of information is required.
b Ecstasy blocks the sites in the brain's synapses where serotonin, a transmitter substance, is removed.
c Conditioned reflexes increase an animal's chance of survival because they enable the animal to learn new things.
d Mammals can adapt to new situations because there is a huge variety of potential pathways in the brain.

B6 GCSE-style questions

1 a She grips a finger that is put into the palm of her hand, she steps when her feet touch a flat surface
b ...something being put into her mouth; milk
c Reflex actions help a worm to respond to stimuli in ways that are most likely to result in their survival, for example in finding food and sheltering from predators.

2 5/6 marks: answer describes several methods that help humans to remember information **and** includes a detailed, logical, and coherent explanation for each method.
All information in the answer is relevant, clear, organised, and presented in a structured and coherent format.
Specialist terms are used appropriately. Few, if any, errors in grammar, punctuation, and spelling.
3/4 marks: answer describes one or two methods that help humans to remember information **and** includes an explanation for each method **but** the explanations lacks detail, logic, and coherence **or** includes an explanation for one method **and** the explanation is detailed, coherent, and logical.
Most of the information is relevant and presented in a structured and coherent format. Specialist terms are usually used correctly. There are occasional errors in grammar, punctuation, and spelling.
1/2 marks: answer describes one method that helps humans to remember information **and** gives an explanation for this method **or** the answer describes one method which helps humans to remember information **but** does not include an explanation for this method.
There may be limited use of specialist terms. Errors of grammar, punctuation, and spelling prevent communication of the science. Answer includes 1 or 2 points of those listed below.
0 marks: insufficient or irrelevant science. Answer not worthy of credit.

Relevant points include:
- Looking for patterns in the information
- This is an example of processing information deeply, which helps humans to remember things
- Repeating the information many times
- This moves information from the short-term memory to the long-term memory
- Attaching a strong stimulus to the information, including colour, light, smell

3 a i Top line 48, bottom line 32
ii Drinking caffeine appears to speed up reaction times.
iii Because it takes time for messages to travel around a reflex arc.
b i The investigation shows that, under the conditions of the test, reaction times are slower if loud music is playing.
ii Kate might not approve of drinking alcohol or Kate might have to drive home or any other sensible answer.

4 a Effector cells – make changes in response to stimulus; receptor cells – detect a stimulus; brain and spinal cord – control the body's response to a stimulus
b i 1 sensory neuron, 2 central nervous system, 3 motor neuron
ii Electrical, peripheral, central

5 The dolphin jumps through a hoop for the first time. A nerve impulse travels along a neuron pathway in the brain for the first time. This makes connections between these neurons. The dolphin jumps through the hoop again. More nerve impulses travel along the same neuron pathway. This makes the connections between the neurons in the pathway stronger. It is now easier for nerve impulses to travel along the pathway, and for the dolphin to jump through hoops.

6 a A nerve impulse arrives at the synapse. A chemical is released from the sensory neuron. The chemical diffuses across the synapse. Molecules of the chemical fit into receptor molecules on the motor neuron. A nerve impulse is stimulated in the motor neuron. The chemical is absorbed back into the sensory neuron to be used again.
b i Cocaine might act in the same way as Ecstasy, blocking the reuptake of serotonin by sensory neurons in synapses.
ii If there is a greater concentration of serotonin in synapses, then the transmission of nerve impulses will increase.

B7.1 Workout

1 True statements: a, d

2 Correct words: synovial, ligament, inelastic, smooth, physiotherapist

3

Component of a joint	Function
cartilage	protects the ends of the bones
tendon	attaches a muscle to a bone
ligament	holds two bones together at a joint

4 Rest, Ice, Compression, Elevation

5 (in any order:)
1 How much exercise do you do already?
2 What is your medical history?
3 Are you taking any medication?

6 Increase, oxygen/glucose, glucose/oxygen, decrease, height, blood pressure

7 Correct sentences: physiotherapy will help an injured joint to heal, but it may take weeks; a fit person will have a lower heart rate than an unfit person; when measuring your pulse, you should not feel it with your thumb.

B7.1 Quickfire

1 Support, movement, protection
2 By contracting and pulling on the bones
3 a Muscles that work against each other
  b Muscles can only contract, so one muscle contracting pulls the bone in one direction and elongates the other muscle, and the other muscle contracting pulls the bone in the other direction.
  c Biceps and triceps
  d Biceps
  e Triceps
4 a To enable growth to occur, broken bones to heal, and bone shapes to be modified
  b Stresses placed on the bones cause them to change shape – they may get thicker.
5 a To allow movement
  b Elbow, knee, fingers
  c Ball and socket
6 a Bone, cartilage, synovial fluid, synovial membrane, tendon, ligament
  b To reduce friction and allow smooth movement
  c To absorb shocks and allow friction free movement
  d Synovial fluid
  e Lubrication – to reduce wear in the joint
7 a A tendon attaches a muscle to a bone.
  b So that muscle contraction is transmitted to the bone, rather than stretching the tendon
8 a A ligament holds bones together at a joint.
  b To allow bones to move
9 a To assess the likelihood of heart disease or lung disease
  b Smoking harms the lungs and reduces the body's ability to absorb oxygen.
  c Certain exercises may need to be avoided to avoid causing further harm.
  d For example, in case you have an asthma attack or similar
10 a Data recorded at start of exercise regime
  b To use for comparison as you get fitter
  c Three from: resting heart rate, blood pressure, recovery after exercise, BMI, proportion of body fat
  d How long it takes for the heart rate to drop to the resting rate after exercise
11 a At the neck or wrist
  b Your thumb has its own pulse.
  c 60–70%
12 a A sphygmomanometer and a cuff around the arm
  b 120/80 mmHg
13 a BMI = mass in kg / (height in metres)^2
  b Measure your height and body mass then calculate the BMI.
  c 18.5–24.9 kg/m^2
  d For example, use calipers to measure the thickness of a skin fold, estimating the amount of fat under the skin.
14 a Sprains, dislocations, torn ligaments, and torn tendons
  b An overstretched ligament
  c Football involves rapid turns and changes of speed.
  d Pain, tenderness, swelling, bruising
  e Rest (or RICE)
  f Ice reduces the swelling.
  g If too tight it could reduce blood flow to the area; if too loose it will not provide sufficient support.
  h It allows gravity to help fluid flow away from the area.
15 a A degree
  b To work out what the problem is and decide what sort of exercise will be the best treatment
  c Movement to free up the stiffness
  d The injury takes a long time to heal and the muscles may take a while to strengthen – repeated exercise helps the muscles to develop in the right way; to strengthen the muscles and ensure full movement in the joint.
  e So that the patient understands why they are important and will continue doing the exercises
16 a How close your reading is to the real value
  b So that you have a true value of body fat or blood pressure and do not act using incorrect figures
17 a Repeatability – whether all the readings for one value are similar
  b Repeat the readings and calculate a mean.
18 Add the readings together and divide by the number of readings.

B7.1 GCSE-style questions

1
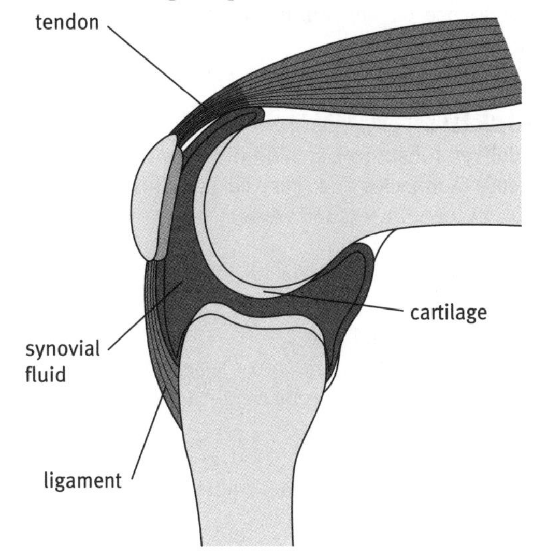

2 a 76 kg/(1.9 m)^2 = 21 kg m^{-2} **b** Normal **c** Data taken at the start of a fitness programme, used to compare with data to show improvement
3 a A pulse meter would be more accurate/easier to use.
  b To ensure the result is repeatable and to calculate a mean
  c A sprain, an overstretched ligament
4 5/6 marks: all information in the answer is relevant, clear, organised and presented in a structured and coherent format. Specialist terms are used appropriately. Few, if any, errors in grammar, punctuation, and spelling. Answer includes 5 or 6 points from those below.
3/4 marks: most of the information is relevant and presented in a structured and coherent format. Specialist terms are usually used correctly. There are occasional errors in grammar, punctuation, and spelling. Answer includes 3 or 4 points of those listed below.
1/2 marks: answer may be simplistic. There may be limited use of specialist terms. Errors of grammar, punctuation, and spelling prevent communication of the science. Answer includes 1 or 2 points from those listed below.
Examples of points to include:
- RICE:
  - Rest to immobilise the injured joint and keep weight off it
  - Ice to reduce swelling
  - Compression using a bandage, to reduce swelling

- – Elevation to allow excess fluid to drain away
- A physiotherapist would assess the injury.
- Then suggest exercises that will ensure flexibility and strength are regained over a period of time.

B7.2 Workout

1 Two, oxygenated, body, oxygen, glucose, carbon dioxide, urea
2 Red blood cells – carry oxygen; white blood cells – fight infections; platelets – help clot blood; plasma – transports dissolved glucose
3 a Haemoglobin binds to oxygen to carry more oxygen than would be possible if it just dissolved in the plasma.
b This leaves space for more haemoglobin inside the cells.
c This provides a large surface area for exchange of gases.
4 Arteries, pulmonary vein, pulmonary artery, left, oxygenated, veins
5 Small, large in number, have thin walls
6 Plasma, glucose, oxygen, amino acids
7 True sentences: A, D, F
8 C, B, D, A

B7.2 Quickfire

1 a To deliver substances around the body (oxygen and glucose to muscles and carry carbon dioxide away)
b Arteries, capillaries, and veins
c Arteries
d Oxygen, glucose
e Vein
f Carbon dioxide and heat
2 a From lungs – pulmonary vein – heart (left side) – aorta – artery – organ – vein – vena cava – heart (right side) – pulmonary artery – lungs
b Double circulation
c In the ventricles of the heart / aorta
d In the veins / vena cava
3 a So that more oxygen can be delivered more quickly
b Mammal can be more active and maintain its metabolic rate.
4 a Plasma, red blood cells, white blood cells, platelets
b They contain a pigment called haemoglobin.
c Oxygen, carbon dioxide, urea, hormones, amino acids
d Oxygen
5 a 5–7 litres
b 5 million
c Approximately 25 000 000 000 000 (2.5×10^{13})
6 a They fight disease.
b White blood cells engulf and digest foreign matter.
c They help to clot blood.
7 a Haemoglobin
b This allows more space for haemoglobin.
8 a A biconcave disc
b It gives a larger surface area to take up and release oxygen.
9 a To pump blood around the body
b 70 bpm
c It goes up.
10 a The atria (singular atrium)
b The ventricles
c The right side
d The left ventricle
e To create the highest pressure, to send blood all the way around the body
f They only need to push blood into the ventricles.
11 a Pulmonary vein **b** Aorta **c** Pulmonary artery
d Vena cava
12 a The atria
b All chambers relax as blood flows into the heart; the pacemaker produces an electrical stimulus; the atria contract pushing blood into the ventricles; the ventricles contract pushing blood out of heart into main arteries.
c The closing of the valves (first the atrioventricular valves, then the semilunar valves)
13 a To ensure blood flows in the correct direction
b These valves prevent blood from flowing up from the ventricles to the atria as the ventricles contract.
c These valves prevent blood from returning from the arteries to the ventricles as the ventricles relax.
d None (but there are valves at the start of the aorta and pulmonary artery where the blood leaves the heart)
e The veins have very low pressure and valves prevent backflow of blood away from the heart down the veins in the legs and arms.
f The contraction of muscles in the legs and arms squeezes the veins flat.
14 a In all the organs between the arteries and veins
b Exchange surfaces
c To provide a very large surface area for exchange
d Artery/arteriole
e Vein/venule
f At the arterial end
g Blood fluid is squeezed out of the capillary to become tissue fluid. The volume of blood in the capillaries therefore decreases, which reduces the pressure.
15 a Small diameter, very thin wall, large numbers
b Red blood cells, platelets, large proteins
c Water, oxygen, nutrients (amino acids or glucose)
d At the arterial end
e Tissue fluid
f Carbon dioxide and water
g At the venous end
h The tissue fluid surrounds the body cells bathing them in oxygen and nutrients. The oxygen and nutrients can diffuse from the fluid through the cell membranes.

B7.2 GCSE-style questions

1 a

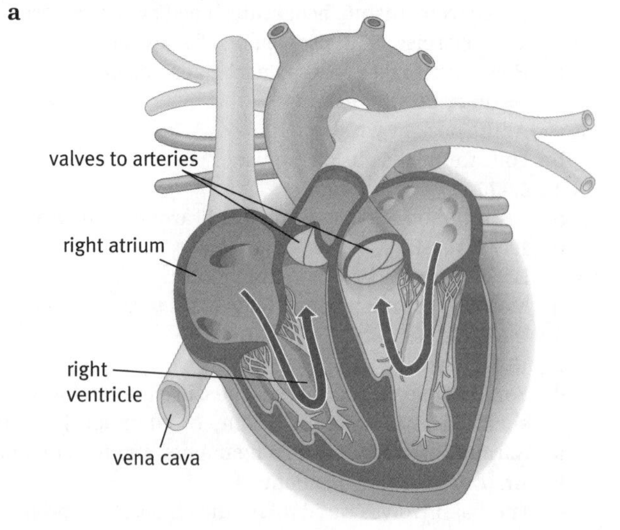

b Via pulmonary vein into left atrium, through atrioventricular valve into left ventricle, through semilunar valve into aorta c Oxygen is picked up by the red blood cells and combines with haemoglobin to produce oxyhaemoglobin.

d Pulmonary artery

2 a

Component of blood	Function
red blood cells	carry oxygen
white blood cells	fight disease
platelets	help clot blood

b To provide a large surface area, for rapid gas exchange

3 a Blood does not need to be at high pressure because the atria only pump blood into the ventricles (a short distance).

b Blood is at low pressure in veins, and valves prevent the backflow of blood.

c The left ventricle must create a higher pressure, to pump blood all around body (a long distance).

4 a Large number of capillaries giving a large surface area; capillary walls permeable to plasma and small particles such as oxygen

b Blood contains **i** red blood cells **ii** white blood cells **iii** platelets; tissue fluid does not.

5 5/6 marks: all information in the answer is relevant, clear, organised and presented in a structured and coherent format. Specialist terms are used appropriately. Few, if any, errors in grammar, punctuation, and spelling. Answer includes 5 or 6 points from those below.

3/4 marks: most of the information is relevant and presented in a structured and coherent format. Specialist terms are usually used correctly. There are occasional errors in grammar, punctuation, and spelling. Answer includes 3 or 4 points of those listed below.

1/2 marks: answer may be simplistic. There may be limited use of specialist terms. Errors of grammar, punctuation, and spelling prevent communication of the science. Answer includes 1 or 2 points from those listed below.

Examples of points to include:

- High pressure of blood at artery end of capillary.
- Capillary wall is porous (allows passage of fluid through it).
- Small molecules can pass through the capillary wall such as oxygen, water, glucose, amino acids.
- These pass into fluid surrounding body cells.
- The tissue fluid carries molecules needed by cells to the cell surfaces.
- This gives a small distance for diffusion into cells.

B7.3 Workout

1 Correct statements: your temperature may be lower at night; your extremities may be cooler than your core; putting more clothes on will increase insulation; sweat cools you down by evaporation.

2 37, respiration, constrict, lose heat / cool down, hypothalamus

3 True statements: A, C, D, E, G, H

4

Type 1 diabetes	Type 2 diabetes
Often appears in young people.	Generally develops later in life.
Cells in pancreas cannot make insulin.	Cells in the body do not respond to insulin
Treated by insulin injections and monitoring diet (carbohydrates).	Treated by careful diet and exercise.

5 Good lifestyle choices: B, E, F, H

6 a Exercising uses energy which was stored as fat.

b The sugars enter her blood and cause the release of insulin. The insulin causes the removal of sugars from the blood leaving less sugars to supply energy to the muscles.

c Fruit and vegetables supply vitamins and other substances (such as antioxidants) which help keep cells healthy. The fibre in fruit and vegetables also helps keep the digestive system healthy.

B7.3 Quickfire

1 a Respiration

b Glucose

c Oxygen

d Carbon dioxide and water

e Exercise increases the respiration rate so more heat is released.

f Muscles (and liver)

2 a 37 °C

b They lose heat to the surroundings more easily.

c In blood flowing round the body

d Skin, brain

3 a An organ that carries out a response

b Sweat glands, muscles

c Your respiration rate is lower at night.

d Heat is lost to the surroundings.

e Put more clothing on, curl up, huddle together

4 a Shivering is contraction of muscles which releases heat.

b Hairs on skin become erect; less blood flows to the skin.

5 a Sweat is produced.

b Sweat evaporates, which requires heat – this heat is taken from the blood.

c Sweat needs energy in the form of heat to evaporate; it takes this heat from the skin. If it does not evaporate then the heat is not removed from the skin.

d If it is humid and there is no breeze the sweat does not evaporate.

6 a Blood sugar levels rise.

b Insulin

c Insulin causes the liver to take sugar out of the blood.

d Tired, weak, irritable

7 a Complex carbohydrates

b For example, pasta, rice, potato

8 a Genetic – the cells of the pancreas cannot make insulin.

b They feel tired and weak; eventually a coma may result.

c Two from: thirst, large volumes of urine, sugar in urine

d Insulin injections and matching carbohydrate intake to energy use

9 a Middle age

b Lack of physical activity, obesity, poor diet

c Regular moderate exercise and low carbohydrate intake

10 a Carbohydrates, proteins, fats, water, vitamins, minerals, and fibre

b For example, obesity, bowel cancer, diabetes, heart disease, anorexia

11 a Exercise, healthy diet, enough sleep

b Exercise uses energy which will not therefore be stored as fat.

c It tones the muscles, improves the efficiency of the heart, and strengthens muscles.

d For example, cycling, walking, running

12 a Being physically active all day uses a lot of energy.

b It will contain less fibre, more salt, and more sugar.

13 a Something that increases the chance of a disease developing

b i High blood pressure ii Obesity iii Too much sugar in the diet iv Not enough fruit and vegetables in the diet v Alcohol
c If a link can be established between a disease and a particular aspect of lifestyle then advice can be given to other people and incidence of the disease can be reduced
d They have a better diet – more fish, less red meat.

14 The hypothalamus

15 a Blood vessels in the skin narrow to reduce blood flow to the skin.
b Less blood flows to the skin surface so less heat is lost to the surroundings.
c Blood is red and causes a pink colour in the skin. Less blood at the skin gives a paler colour.
d The blood vessels in the skin dilate. More blood flows to the skin surface and gives a red colour.

16 a Working against each other
b Vasodilation that brings more blood to the skin surface and vasoconstriction that reduces blood flow to the skin surface
c They provide a sensitive mechanism with fine control.

B7.3 GCSE-style questions

1 a Swimming
b 29.3 / 14.2 = 2.1
c A domestic cleaner is being more active, which uses energy, 11.3 compared with 1.7 kJ/h.

2 a The hypothalamus detects the blood temperature, processes the information, and triggers effectors to correct it if it lies outside the normal range.
b The hairs stand on end; less sweat is made; vasoconstriction reduces blood flow to skin.
c The effectors work against one another to give a sensitive mechanism and fine control.

3 a i Pancreas ii A high blood sugar level iii Insulin causes cells to take up sugars from blood, and reduces blood sugar levels.
b i They cannot make their own insulin, so need insulin from another source to regulate blood sugar level.
ii Regular exercise uses sugars in the blood, so prevents the level from getting too high.

4 a Eating lentils reduces the risk of developing bowel cancer by a third.
b Fibre / bran
c Green vegetables contain fibre which helps move food through the gut. They also provide vitamins to keep cells healthy.

5 5/6 marks: all information in the answer is relevant, clear, organised and presented in a structured and coherent format. Specialist terms are used appropriately. Few, if any, errors in grammar, punctuation, and spelling. Answer includes 5 or 6 points from those below.
3/4 marks: most of the information is relevant and presented in a structured and coherent format. Specialist terms are usually used correctly. There are occasional errors in grammar, punctuation, and spelling. Answer includes 3 or 4 points of those listed below.
1/2 marks: answer may be simplistic. There may be limited use of specialist terms. Errors of grammar, punctuation, and spelling prevent communication of the science. Answer includes 1 or 2 points from those listed below.
Examples of points to include:
Regular exercise:
- uses excess energy
- keeps body weight down
- keeps joints and muscles moving
- maintains flexibility and strengthens the muscles.

A healthy diet:
- provides nutrients for respiration and growth
- keeps the cells healthy
- supplies energy for activity.

At least two points should be made for exercise and two for diet.

B7.4 Workout

1 Sustainable, substrate, bacteria, enzymes, erosion

2 True statements are: human systems are linear. Earthworms help to recycle mineral nutrients. Deforestation causes flooding. Oil is fossil sunlight energy.

3 D, A, B, C, E

4 a The loss of foliage to protect and enrich soil and the loss of roots to hold soil together allow soil erosion. Loss of soil can cause desertification.
b There is little or no input or loss of nutrients. Wastes from one process are recycled or reused in another process.
c Open-loop systems allow loss of wastes and require inputs.
d Bacteria cause decay which releases nutrients to the soil for plants to use.

5 Biofuel – fuel made from plant growth; fossil sunlight energy – energy stored millions of years ago; intensive agriculture – a human system involving high inputs; open-loop system – a system with a lot of waste; quota – a limit to the number of fish that can be caught

6 a A life support system supplied by an ecosystem that is essential for our survival
b For example, a supply of clean water, removal of carbon dioxide from the air, pollination of crops

7 a This increases the chances of fertilisation and of survival of the species.
b Insects pollinate many flowering plants; insects may also eat pests that damage crops.
c The system has developed with few losses (as wastes are reused) and can survive with few or no inputs other than energy.
d Rivers and streams can carry away some materials such as leaves; animals may feed in one ecosystem and migrate to another.
e Humans often harvest materials and remove them from the ecosystem – this removes materials that cannot then be reused and recycled within the ecosystem.

8 Correct statements: vegetation protects the soil from erosion; agriculture reduces biodiversity; dead organic matter contains energy; mineral nutrients are recycled by bacteria.

B7.4 Quickfire

1 a A system with many inputs and many losses – little is recycled.
b Intensive farming with a high input of fertiliser and fossil fuel; the products are removed to be eaten.
c They are dumped.
d A system in which there is little input and the wastes from one process become the substrate for another process.
e A natural ecosystem in which plants remove nutrients from the soil and return them when the leaves drop off and decay
f The wastes from another process within the system
g Open loop

2 a A process which can be maintained without damage to the environment

- **b** They take too much material from the environment, or add too much waste to the environment.
- **c** There are losses due to wind or water carrying materials away.
- **d** For example, leaves blown on the wind; seeds carried away by birds
- **e** So that there are no overall losses or gains and the system is sustainable.

3 a The sun
- **b** Carbon dioxide is taken in by plants for photosynthesis; it is returned to the atmosphere by respiration in both plants and animals.
- **c** It returns carbon dioxide to the atmosphere that was absorbed millions of year ago.
- **d** Nitrates are taken up by plant roots and converted to proteins. Animals eat the plants and excrete nitrogen as urea which is broken down to nitrates again by bacteria. Bacteria also convert the protein in dead organic matter to nitrates. Some bacteria can also fix nitrogen from the air.

4 a Microorganisms use the wastes and organic matter as a substrate and release nutrients back to the soil.
- **b** Bacteria or fungi
- **c** Bacteria release digestive enzymes which break down dead organic matter to mineral nutrients.

5 a It explains how a closed-loop system could work – after materials are taken and used they are reused and recycled.
- **b** It means that less inputs are required and less waste is disposed of.
- **c** Pollen is released in huge numbers to try to make sure that pollination occurs and the plant species reproduces successfully to survive.

6 a The removal of forests to leave clear ground
- **b** Foliage protects soil from erosion by rainfall; the soil with plant roots growing in it can absorb water and prevent flooding. The water may be released more slowly as it drains through the soil. Forests help to recycle water by increasing evaporation from the ground to produce clouds and rainfall.
- **c** Loss of foliage exposes soil to erosion by heavy rain. The roots are lost and no longer bind soil together so it can be washed away.
- **d** Rivers and lakes may become silted up as soil is damaged and washed away.

7 a Biodegradable means that the material can decompose naturally.
- **b** For example, plastic and glass

8 a A life support system that we depend upon for our survival
- **b** For example, the production of clean water from rainfall; the supply of oxygen from plant photosynthesis
- **c** Humans removing too much material from an ecosystem making it unstable

9 a By taking too many fish (overfishing)
- **b** Fish stocks will run out.

10 a The plant removes the nutrients from the soil and the soil loses its fertility. Organic matter is not returned to the soil and the soil structure is damaged. Eventually the soil will dry out and become eroded.
- **b** They remove nutrients from the soil by eating vegetation; they compact the soil preventing drainage; they trample plants and reduce biodiversity.
- **c** Herbicides kill weeds, and livestock trample plants. This limits the number of species that can survive and limits the number of insects and other herbivores living in the ecosystem.
- **d** A range of plants on the farm will provide a wider range of habitats for insects and other animals. These may be pollinators or they may eat pests that damage the crop.
- **e** Excess fertiliser gets washed into rivers causing algal blooms and eutrophication.

11 a Millions of years
- **b** Incompletely decomposed organic matter
- **c** The Sun
- **d** The wastes (carbon dioxide and other gases) are additional inputs to the system.

12 a A set amount of material (such as fish) that can be harvested
- **b** To prevent overfishing which will deplete fish stocks and unbalance the ecosystem
- **c** The seas are boundaries between countries and if one country limits its harvest it will be of no use if other countries simply take more.
- **d** Breeding fish on a farm and releasing them into the wild
- **e** They will have had the opportunity to reproduce, ensuring that the species can survive.

13 a Selective felling (selecting certain types and sizes of trees), replanting, legislation to prevent clear felling
- **b** Bacteria that live in the root nodules of legumes and take nitrogen out of the air to produce nitrates and proteins
- **c** Growing legumes such as clover or beans in the fields and ploughing the dead plant roots back into the soil
- **d** Different crops are grown in a field in successive seasons. Each crop removes different nutrients from the soil. Some crops may add nutrients to the soil; other crops may improve soil structure. Rotation also stops pests from synchronising with the crop and causing a lot of damage.
- **e** Adding natural manure or compost to the soil, avoiding ploughing too often

14 a For example, grow plants and eat or burn the produce; convert the energy to electricity; use the energy to heat water in solar panels.
- **b** The Sun will not fail (for the foreseeable future).

15 a For example, take small amounts of produce, return all materials to the ecosystem.
- **b** The population gets too large and removes too many materials from the ecosystem; they may produce too many wastes for the ecosystem to cope with.
- **c** All species have a right to live and survive. The ecosystem may be affected by the loss of a species. It is important for humans to maintain biodiversity because the natural world may provide materials and technologies that will be useful to us in the future.
- **d** There may be unforeseen effects of losing an ecosystem – it could destabilise another ecosystem or impact upon our production of food.

16 a The concentration of a chemical (toxin) becomes more concentrated higher up a food chain.
- **b** The toxic substances are found at lower concentrations lower down the food chain.
- **c** If plants contain a low concentration of toxin, the primary consumers that eat several plants accumulate a higher concentration. In a secondary consumer that eats many primary consumers, the concentration increases further, and so on.

17 a Extra nutrients in a water body cause algal growth. This blocks the light and kills the plants. Bacteria act to decompose the dead plants, using up all the oxygen and killing animals.

- **b** Increased growth of algae
- **c** Decay and decomposition
- **d** Bacteria
- **e** Lack of oxygen

B7.4 GCSE-style questions

1 **a** A closed-loop system
b There is little or no waste; the wastes from one process are reused as inputs for another process.
c Microorganisms recycle minerals and nutrients. They do this by decomposing dead organic matter by releasing enzymes which digest the dead organic matter.

2 5/6 marks: all information in the answer is relevant, clear, organised and presented in a structured and coherent format. Specialist terms are used appropriately. Few, if any, errors in grammar, punctuation, and spelling. Answer includes 5 or 6 points from those below.
3/4 marks: most of the information is relevant and presented in a structured and coherent format. Specialist terms are usually used correctly. There are occasional errors in grammar, punctuation, and spelling. Answer includes 3 or 4 points of those listed below.
1/2 marks: answer may be simplistic. There may be limited use of specialist terms. Errors of grammar, punctuation, and spelling prevent communication of the science. Answer includes 1 or 2 points from those listed below.
Examples of points to include:
- Loss of vegetation exposes soil to rainfall.
- Loss of roots means the soil is no longer bound together.
- Soil will be eroded by rainfall and washed away.
- This may cause silting of rivers and lakes.
- Flooding is possible.
- Less water is evaporated from leaves.
- Less clouds and rainfall result.
- Drought may result.

3 **a** Agriculture that involves high inputs and high yields
b The crop plants remove nutrients from the soil. These must be replaced or the soil becomes infertile.
c For example, herbicides, fossil fuels

4 **a** Bees pollinate plant flowers; bees also make honey.
b The leaves protect the soil from heavy rain; the roots bind the soil together.
c Forests enable the evaporation of water to create clouds and rainfall. The soil can act as a sponge and holds water so that flooding does not occur. The water slowly drains through the soil providing water over a period of time after the rain – this prevents droughts.

5 Extra nutrients cause algal growth in a body of water. This blocks light, so plants die. They are decomposed by bacteria, which use up all the oxygen. This results in the death of animals.

B7.5 Workout

1 Quickly, enzymes, fermenters, vector, 100
2 C, B, D, A
3 **a** Chymosin is produced by fungi so it is not derived from animals.
b Stem cells are not specialised. When they develop they can differentiate into a number of different types of cell.
c The human gene for insulin can be inserted into bacteria which then incorporate the gene into their own DNA and use it to code for the proteins they make.
d Biological washing powders contain enzymes which digest the dirt on clothing.

4 True statements: heart valves can be replaced; new skin can be grown using tissue culture; fungi can be used to make medicines; bone marrow contains stem cells.
5 1 differentiation; 2 bacteria; 3 probe; 4 chymosin; 5 nanotechnology; 6 fungus; 7 antibiotic; 8 enzyme; 9 vector
6 David

B7.5 Quickfire

1 **a** Microscopic living things, including bacteria and fungi
b For example, they have no true nucleus, they are smaller.
c They can be used to manufacture biological molecules such as enzymes.
d They grow and reproduce quickly; they possess plasmids; they have simple biochemistry; they can make complex molecules; there are no ethical concerns over their use.
e Fungi have proper nuclei and more complex biochemistry; not all are single celled.

2 **a** Growing microorganisms in fermenters under ideal conditions
b Antibiotics, insulin, single cell protein, chymosin
c Temperature, nutrient levels, and waste levels

3 **a** A protein made by fungi and used for food.
b Fungi
c It can be processed for human food, e.g. for vegetarians, or for animal feed.

4 **a** Calf stomachs
b To make milk proteins coagulate for making cheese
c Genetically engineered bacteria
d The chymosin does not come from calves.

5 **a** Fuels made from plants that have been grown recently
b Yeast is used to ferment the sugar anaerobically.
c Growing crops for biofuels might use land that was being used for producing food.
d Lignocellulase

6 **a** A washing powder that contains enzymes
b Food stains – grease (fat), carbohydrate, or protein

7 **a** Altering the DNA of a microorganism by adding genes from another species
b Plasmids
c It needs the correct code in the sequence of DNA bases for the human protein.
d Insulin
e A method of transferring a gene into another organism
f They occur in bacteria and are small enough for bacteria to absorb as they grow.
g The gene is isolated and inserted into plasmids. The plasmids are mixed with bacteria, which absorb the plasmids.

8 **a** For example, herbicide resistance
b Farmers can spray their crop to kill weeds that compete with the crop without harming the crop itself.
c Some people think it is 'playing God' – interfering with nature. There may be unexpected results in the recipient organism. There is the potential to upset the natural balance if modified organisms breed in the wild.
d Selective breeding is also interfering with nature and most people have no no ethical objections to this. Research is very thorough and genetically modified organisms are carefully monitored and checked. GM plants are not allowed in areas where they may interbreed with wild plants.

9 **a** Very small
b 100 nm or smaller

c To inhibit the growth of bacteria, to indicate oxidisation when packages are leaking, to supply antibodies to react with bacteria
d 1000 000 000 (one thousand million)
e Silver

10 a A cell that has developed in a specialised way
b To enable them to do a specific job well
c A cell that is capable of dividing and differentiating into a range of cell types
d In bone marrow, in umbilical cord blood

11 a Growing tissues from cells in the laboratory
b To treat severe scars or burns
c Cells from the bone marrow of a healthy person are injected into the bone of a person with leukaemia.
d To replace pancreas cells in people with diabetes, to repair nervous tissue in people with paralysis, treating people with Alzheimer's disease
e The stem cells must be collected from a healthy person.
f The stem cells must be a similar type or they may be rejected by the immune system.

12 a The use of modern technology to replace human body parts
b Valves, pacemaker
c To set the rhythm of the heart beat
d If the heart keeps going into an unusual rhythm
e To ensure the blood flows the correct way
f If they leak badly and allow blood to flow in the wrong direction
g From a human or other animal donor; an artificial valve
h Resistant to wear and tear; unreactive so they do not corrode; do not stimulate the body's immune system

13 a Two chains of subunits wound into a double helix; the chains are held together by pairs of the bases adenine, thymine, cytosine, and guanine.
b The sequence of the bases provides a code for the characteristic; the bases pair in a specific way – adenine to thymine and cytosine to guanine.
c Blood
d They have no nucleus.

14 a A length of single-stranded DNA manufactured with a specific sequence complementary to the gene being looked for
b The heat separates the two chains, exposing the bases on each chain so that the probe can bind
c DNA is not visible – the marker fluoresces to make the probe visible.
d Using gel electrophoresis
e For example, cystic fibrosis, Parkinson's disease

B7.5 GCSE-style questions

1 a Bacteria and fungi
b For example, antibiotics and insulin

2 a i The rate of fermentation is affected by temperature and by enzymes. I predict that as temperature increases the rate of fermentation increases up to a certain temperature; above that temperature the rate of fermentation falls as temperature increases further.
ii Higher temperatures give the molecules more kinetic energy. However, once the temperature rises too high the enzymes are denatured and stop working.
b i A fungus (yeast)
ii A stopclock and a thermometer
c i The results match the first part of my prediction and back up the first part of the hypothesis. As the temperature rises from 10 °C to 40 °C the rate rises from 0.002 to 0.014 s^{-1}. Temperature does affect the rate of fermentation. However, the results do not show that rate of fermentation drops at higher temperatures.
ii She could repeat the experiment more times at each temperature.

3 a Production can easily be increased to meet demand; the insulin is human insulin and acts exactly as if it had been made in the pancreas.
b The production does not involve calves and is therefore suitable for vegetarians; there are no ethical issues over growing bacteria to produce chymosin.
c Production is very rapid; the protein is virtually fat free; it can be used as a meat substitute for vegetarians.
d The nanoparticles can keep the food fresher; they indicate when the wrapping is leaking or when the food is going off.

4 5/6 marks: all information in the answer is relevant, clear, organised and presented in a structured and coherent format. Specialist terms are used appropriately. Few, if any, errors in grammar, punctuation, and spelling. Answer includes 5 or 6 points from those below.
3/4 marks: most of the information is relevant and presented in a structured and coherent format. Specialist terms are usually used correctly. There are occasional errors in grammar, punctuation, and spelling. Answer includes 3 or 4 points of those listed below.
1/2 marks: answer may be simplistic. There may be limited use of specialist terms. Errors of grammar, punctuation, and spelling prevent communication of the science. Answer includes 1 or 2 points from those listed below.
Examples of points to include (must be in the correct sequence):
- The gene for human insulin is isolated.
- Plasmids are cut open.
- The isolated gene is inserted into the plasmids.
- The plasmid is used as a vector.
- The plasmids are mixed into a medium to grow bacteria.
- Bacteria are grown in the medium.
- The bacteria take up the plasmids.
- The bacteria are grown in large numbers in a fermenter.
- They produce the human protein which can be harvested.

5 a One
b Shining light that makes it fluoresce, or UV light
c The probe is used to identify a certain genetic sequence; it is mixed with DNA that has been treated to make it single stranded.

Ideas about science 1 Workout

1 Factors he should consider: he should take repeat readings; a fit person will have a lower heart rate; the time of day when his pulse is counted; he should calculate a mean.
2 Repeatable, incorrect, true, verify

Ideas about science 1 GCSE-style questions

1 a Lilia could do the whole investigation twice more, recording the number of bubbles produced in 1 minute at each temperature. If the numbers of bubbles in the three investigations are similar, then her results are repeatable.
b Lilia could ask someone else in the class to repeat the investigation using a different set of equipment. If the other person obtains similar data (the change in the number of bubbles with temperature shows a similar pattern) then Lilia's results are reproducible.

2 a i A single reading could be inaccurate; taking three readings provides a range so that we can compare the readings and look for repeatability; a mean gives a figure we can be more confident is close to the true value.
ii Not use his thumb to count the pulse
b i His resting pulse rate gets lower; his heart rate after exercise is also lower. Both these measurements suggest that his fitness has improved.
ii In the first table there is a correlation between time from start of program and a lower pulse rate. There is an unexpected result at day 21: looking at the raw data for day 21 there is one reading that may be incorrect – the third count. This means that we have less confidence in the data for day 21. We have more confidence in the data for day 7 as all the results are close.

Ideas about science 2 Workout

1 True statements: there is a correlation between BMI and recovery time; as BMI increases, recovery time also increases.
2 A2 larger, B3 correlation, C4 chances, D1 control

Ideas about science 2 GCSE-style questions

1 a The masses of ester compounds
b i Amount of water; temperature
ii So that the investigation is fair
c i As the concentration of carbon dioxide in the air increases, the concentration of ester compound A in the strawberries increases; as the concentration of carbon dioxide in the air increases, the concentration of ester compound B in the strawberries increases.
ii Kezi and Sahira
d The scientist suggested a mechanism for the way in which the extra carbon dioxide resulted in higher concentrations of esters A and B. A plausible mechanism linking a factor to an outcome makes scientists more likely to accept that there is a causal relationship between the factor and the outcome.
2 a Outcomes: of the people who took aspirin, 10 got cancer; of the people who took aspirin placebo tables, 23 got cancer. Factors: for 2 years, 258 people took aspirin every day; for 2 years, 250 people took aspirin placebo tablets every day.
b There is a correlation between the factors and the outcome; taking aspirin reduces the chance of getting bowel cancer.
c So that other factors are equally likely in both groups.
d The suggested mechanism increases confidence that taking aspirin causes a reduced chance of bowel cancer; it is possible that the decreased chance of getting cancer after taking aspirin for 2 years is caused by some other factor.
e i Smoking, drinking alcohol, any other sensible suggestions
ii One of: the size of the sample was very big; the study continued for a long time period (20 years).
3 a That pondweed bubbles when in light
b In brighter light there will be more bubbles.
c Brighter light provides more energy for photosynthesis; more photosynthesis will produce more oxygen gas.
d i No
ii In dim light plants do not photosynthesise; they need bright light to make photosynthesis occur fast enough to produce oxygen.
iii Increase the intensity of light by using several lamps

Ideas about science 3 Workout

1 1 D; 2 A; 3 E; 4 B; 5 C
2 1 A; 2 E; 3 B; 4 D; 5 C
3 1 A, 2 D, 3 E, 4 B, 5 C
4 Correct order: 1E, 2B, 3F, 4G, 5A, 6D, 7C
5 a Science tries to explain observations. If a prediction is made it can be tested against the real observations.
b Data from experiments give us measureable values that can be compared against the predictions.
c If the data from an experiment matches the prediction it gives us confidence that the explanation is correct.
d If the observations do not agree with the predictions, one or other must be incorrect.

Ideas about science 3 GCSE-style questions

1 Data: B, C, E, F; explanation: D
2 5/6 marks: answer clearly identifies all the stages of developing a scientific explanation in the article **and** clearly shows how the article exemplifies each stage. All information in the answer is relevant, clear, organised, and presented in a structured and coherent format. Specialist terms are used appropriately. Few, if any, errors in grammar, punctuation, and spelling.
3/4 marks: answer identifies some of the stages of developing a scientific explanation in the article **and** shows how the article exemplifies each of the stages identified **or** answer identifies all of the stages of developing a scientific explanation in the article **but** does not show how the article exemplifies each of the stages identified.
Most of the information is relevant and presented in a structured and coherent format. Specialist terms are usually used correctly. There are occasional errors in grammar, punctuation, and spelling.
1/2 marks: answer identifies one or two of the stages of developing a scientific explanation in the article. **and** shows how the article exemplifies this stage. There may be limited use of specialist terms. Errors of grammar, punctuation, and spelling prevent communication of the science. Answer includes 1 or 2 points of those listed below.
0 marks: insufficient or irrelevant science. Answer not worthy of credit.
Relevant points include:

- The first paragraph describes data the scientists collected by observation – there were more species of fanged frogs on the small island, and there were no *Platymantis* frogs.
- The second paragraph describes a hypothesis, which accounts for the data and which was developed using both data and creative thought.
- From this hypothesis, the scientists made a prediction – that every type of habitat on the small island would have its own species of fanged frogs.
- The scientists then collected data to test the prediction, by identifying fanged frog species in the different habitat types of the small island.
- The scientists' prediction was correct, since many habitats had their own species of fanged frog.
- The fact that the prediction was correct increased the scientists' confidence in their explanation (the hypothesis).

3 a A: C; B: D; C: D; D: E; E: D
  b The data increase confidence in the explanation; the data agree with the prediction.
4 The data supports the prediction, which is an indication that it would be reasonable to accept the explanation. However, the values for blood pressure change are very similar for all four groups, which is perhaps a reason to be sceptical of the explanation.

Ideas about science 4 Workout

1 a C b X c X
  d C e X

Ideas about science 4 GCSE-style questions

1 a Four from A, B, D, F, I, J
  b C, E, G
  c H, K
2 a i To check their methods and data for obvious problems, as part of the peer-review process
    ii So that other scientists can try to reproduce their findings, and so that other scientists can build on their findings and plan further research in the same area
  b The new data disagreed with predictions from the previous explanation.
  c Their findings make other scientists more likely to accept the claim of the 2005 scientists.
  d Wine-making company – just two studies show that moderate drinking does not reduce the risk of heart disease. Many more studies show the opposite. People can continue drinking as usual
    Organisation that persuades people not to drink alcohol – the 2005 study shows that the results of the earlier research could well be wrong. People who drink alcohol are not protecting themselves against heart disease.
    European health organisation – we need more information before we can assess whether the claim of the 2005 research is correct.
3 a To plan how to find out if the other scientists' work is reproducible; to find out what is already known about the effects of exercise and BMI on hip fracture risk.
  b The Million Women Study research reproduced some of the findings of earlier research.
  c The paper is submitted to a scientific journal. Then a few other scientists, who are experts in the same area of science, check the methods and findings. The paper may then be accepted for publication.
  d To let women know that exercising can reduce their risk of hip fractures.
4 a In a journal
  b Statement C
  c There are no data given to substantiate the claims; it is new technology and new research; it is the first time that scientists have reported being able to grow complete organs; the tissues and organs produced have not been trialled in humans; they have not existed long enough to check they are normal and won't continue growing or become cancerous.
  d To check that the procedures work as claimed, to check that the same results can be achieved.
  e The cells were extracted from a specially created human embryo.

Ideas about science 5 Workout

1 Statements that represent risks: a new herbicide may cause cancer; modified genes may escape to the wild population; no one knows what effect nanoparticles may have; a new exercise may cause a muscle strain; transfats used in baking are new man-made fats.
2 a How can we assess the size of a risk?
  b What do we need to take into account when making a decision about a particular risk?
  c How might we make a decision about a particular course of action that is known to have risks?
  d What makes someone more willing to accept a particular risk?
  e Why might someone be more willing to accept the risks of riding a bike than of flying in an aeroplane?
  f What types of risks are people more willing to take – actions that have long-term effects, or actions that have short-term effects?
  g Sometimes people overestimate, or underestimate, the risk of a particular activity. What is the name given to the risk that people imagine an activity has?
  h In general, when people take part in a new activity, are they more likely to underestimate, or overestimate the risk?
3 1 harm; 2 chance; 3 advance; 4 ionising; 5 consequences; 6 assess; 7 controversial; 8 statistically; 9 benefit; 10 perceive; 11 acceptable
4 **Sentences might include the following:**
  - People's perceptions of the size of a particular risk may be different from the statistically estimated risk.
  - The perceived risk of an unfamiliar activity is often greater than the perceived risk of a more familiar activity.
  - Governments have to assess what level of risk is acceptable in a particular situation. This decision may be controversial.

Ideas about science 5 GCSE-style questions

1 a It makes some users feel happy for up to three hours; peer pressure
  b He has chosen to take the drug himself; he perceives the risks as being short-term only.
  c Its use only became widespread in 2007.
  d Sweating and headache (study A); agitation (study B)
2 a Risks – cannabis may lead to anxiety and panic, it slows reaction times, and leads to coordination and memory problems. Benefits – studies in humans and mice show that cannabis may reduce stiffness and trembling.
    The answer should include an overall assessment of the risk, backed up with evidence.
  b People may think it is ethically wrong to withhold a possible treatment from people with MS.
  c i Will, Yasmin
    ii Verity, Will, Xena, Yasmin

Ideas about science 6 Workout

1 Notes might include the following:
  - Title: Ethical issues in science
  - Most important points: some questions, including ethical ones, cannot be addressed using a scientific approach; ethics is a set of principles that may show what decision to make in a particular situation.
  - Other information: an example of an ethical issue is whether or not to use embryonic stem cells to treat disease; there are different types of ethical arguments, including 'the right decision is the one that gives the best outcome for the greatest number of people' and 'some actions are always right or wrong, whatever the consequences'.
2 All 11–12-year-old girls should be vaccinated against cervical cancer – This decision is the one that benefits

the most people. No one should be vaccinated against cervical cancer. It interferes with nature – Things that are unnatural are never right. 1340 out of 700 000 girls given the cervical cancer vaccine complained of side-effects. Most were minor effects such as rashes and a feeling of dizziness – This statement assesses the risk of having the cervical cancer vaccine. 1300 girls suffer side-effects. Doctors believe 700 lives could be saved in the future – This statement compares the risk with the potential benefit.

3 Questions that a scientist could try to answer: Can nuclear power meet our energy needs? Can nanotechnology keep food fresh? Is organic food more nutritious? Can intensive farming grow more food? Is it possible to develop a malaria vaccine?

Ideas about science 6 GCSE-style questions

1 **a** Agree: Marcus, Nikki; disagree: Kirsty, Oliver, Linda

b Scientific approach: does the technique damage embryos? Do embryos that have been tested grow properly? Is PGS necessary – maybe embryos can fix their own genetic defects?
Values: is PGS ethically acceptable? Is it natural to choose which embryo to implant? Is it right to destroy embryos that are not implanted?

c Different countries have different laws relating to the ethics of the process; some countries decide not to spend money on treatment.

2 **a** Arguments for: reduces number of breast cancer deaths. Arguments against: can lead to unnecessary treatments which may have harmful side effects.

b Most breast cancer cases are in women aged over 50. Since the life expectancy in Swaziland is just 32, relatively few people can expect to get breast cancer there. The percentage of people with HIV/AIDS is much greater in Swaziland than in the UK, suggesting that resources would be better focused on HIV/AIDS in Swaziland.

3 **a** There are risks associated with the side effects of all vaccines.

b A big proportion of the population – or perhaps the whole population – would be protected from mumps, measles, and rubella if the MMR vaccine was compulsory.

4 5/6 marks: all information in the answer is relevant, clear, organised, and presented in a structured and coherent format. Specialist terms are used appropriately. Few, if any, errors in grammar, punctuation, and spelling. Answer includes 5 or 6 points from those below.
3/4 marks: most of the information is relevant and presented in a structured and coherent format. Specialist terms are usually used correctly. There are occasional errors in grammar, punctuation, and spelling. Answer includes 3 or 4 points of those listed below.
1/2 marks: answer may be simplistic. There may be limited use of specialist terms. Errors of grammar, punctuation, and spelling prevent communication of the science. Answer includes 1 or 2 points from those listed below.
Examples of points to include:

- Population in UK is not too large.
- Sufficient food can be grown on the land available.
- Organic food is more expensive to produce.
- People are affluent enough to pay for food.
- In India the population is very large.
- It is not easy to grow enough food to feed everyone.
- Intensive farming can produce more food on land available.
- Some people are living in poverty and cannot afford more expensive food.